BLOCKCHAIN-TECHNOLOGY IN THE ENERGY INDUSTRY

Blockchain as a driver of the energy revolution?

Foreword

When you read blockchain, you usually think of crypto currencies. But there is much more to this technology. In the future, it could revolutionize the entire energy industry as a "digital driver for the energy revolution". Due to environmental pollution and climate change, an **energy transition** is necessary, which is primarily based on a **decentralized energy supply** with **renewable energies** and ensures a sustainable energy supply through the **digitalization of the energy transition** and the use of associated new technologies such as the blockchain. It is therefore all the more important to understand exactly what the advantages of the blockchain are and how they can be used to transform the energy system. This paper pursues exactly this approach and offers a complete overview from the beginnings of the energy system transformation, over a detailed description of the functionality of the blockchain technology up to a concept for a concrete application in the energy industry. By means of an understandable description, the work also aims to make the benefits of blockchain technology - especially in the field of energy management - accessible to readers from outside the field. In the following only the male form is used for reasons of better readability. However, both male and female persons can be meant.

Bartek Mika

Abstract

The expansion of renewable energy is rapidly increasing as part of the energy revolution. The structure of energy supply systems is becoming increasingly decentralized (decentralization). New players, such as prosumers, who generate and consume their own electricity, could establish themselves in the electricity market. However, due to their low capacity, prosumers are currently unable to participate economically in electricity trading. In particular, the increasing complexity of control and the load on the network infrastructure as well as the high requirements on data security, which are associated with the exchange of electricity and the associated electricity bills, require digitalization of the energy revolution (Energiewende 2.0). The aim of this work is to examine if the "*blockchain as a driver of the energy revolution*" for the development of new digital business models can contribute to the successful transformation of the energy system. Numerous statements from energy industry experts, study results and two surveys indicate that blockchain has high potential in the medium and long term to significantly impact the energy industry in the coming years. Blockchain technology promises economic value through its strengths such as disintermediation, security, transparency and automation. However, in addition to technical challenges such as the upcoming smart meter rollout, the smart meter gateway required for communication, and the compatibility between the smart metering systems and the blockchain, there are also legal and regulatory hurdles which make the use of the blockchain difficult in the short-term. By far the most widely discussed usage of blockchain in the energy sector is the peer-to-peer trading of decentralized electricity from renewable energies. Therefore, it was examined within the framework of a concept, whether there is a possibility for prosumers to participate economically in electricity trading, despite their low capacity. The results show that due to regulatory constraints such an implementation is only possible in form of a service model in which all areas of responsibility are transferred to a service provider (eg electricity supplier). An independently developed business model, which includes peer-to-peer trading based on a service, shows the required infrastructure, a detailed process description in the context of a business process map and one option to configure the blockchain.

Keywords: Energy Revolution, Decentralization, Prosumer, Digitalization, Blockchain-Technology, Peer-to-Peer

Table of Contents

1 Introduction 4
2 Decentralised Energy Supply 5
   2.1 Introduction 5
   2.2 Definition and Characteristics 6
   2.3 Benefits for the End User 6
   2.4 Challenges for Energy Suppliers and Distribution System Operators 8
   2.5 Development Goals and Future Prospects 8
3 Digitalization of the Energy System Transition - Energy System Transition 2.0 11
   3.1 Introduction 11
   3.2 Smart-Energy 12
   3.3 The Challenges of Digitalization 16
   3.4 Development Goals and Future Prospects 18
4 Basic Principles of Blockchain Technology 21
   4.1 Introduction 21
   4.2 Functionality of a Blockchain 24
   4.3 Classification of Blockchains 32
   4.4 Strengths of Blockchain Technology 33
   4.5 Weaknesses of Blockchain Technology 36
5 Blockchain in the Energy Industry 39
   5.1 Assessments by Experts from the Energy Industry 39
   5.2 Opportunities and Challenges 48
   5.3 Use Cases of Blockchain Technology in the Energy Sector 51
6 Concept for P2P Electricity Trading via Blockchain 52
   6.1 Introduction 52
   6.2 Problem Formulation 53
   6.3 Objectives and Questions 54
   6.4 Evaluation of potential 55
   6.5 Business Model 60
   6.6 Key Findings and Recommendations for Action 70
7 Conclusion and Outlook 71
Literature 75
Appendix 100

1 Introduction

We live at a time when energy systems are undergoing major changes. Within the framework of the "Energy transition", comprehensive, politically motivated targets were formulated for the development of the energy supply structure. In addition to phasing out nuclear power and expanding renewable energies (RE), the most important economic and environmental policy tasks are to increase energy efficiency (cf. Radtke and Kersting 2018, p. 94). The increasingly decentralised generation structure requires a new approach to the exchange of digital information and a reorganisation of customer relationships. More and more passive consumers are developing into prosumers who actively participate in the design of the power supply system. "*Prosumers are becoming increasingly important for the energy transition*" (BMWi, 2016). This change is a further step on the way to the emancipation of energy customers from the established energy supply companies (EVU), which will result in changes in the supply of energy and accompanying services (Servatius et al. 2012, p. 288f.). This will create a variety of challenges, but also opportunities, for the transformation of energy systems. Energy suppliers are currently confronted with increasing competition and falling yields in the electricity market. In particular, the sheer volume of generation and consumption units and their intelligent balancing, as well as the increasing number of prosumers, illustrate the necessary use of digital technologies. "*A successful energy transition is inconceivable without comprehensive digitalization*" (Neugebauer 2018, p. 347). At the same time, the digital energy revolution is giving rise to new problems such as the secure and protected collection, storage, transmission and processing of data (Richard et al. 2019, p. 10). Although the prosumers make an important contribution to energy system transformation, they are currently not represented in the actual market, since the marketing of electricity continues to be in the hands of aggregators such as the grid operator. The processing of all transactions via the classical energy providers represents a significant barrier to market entry for small producers with low services because the transaction costs are too high in relation to their transaction value due to administrative and regulatory hurdles (Voshmgir 2016, p. 24). As a result, the future energy market is threatened with the loss of a decisive player due to bureaucratic barriers (Glatz 2018, p. 7f.). The overall objective of this paper is to find out whether blockchain technology with its positive characteristics as a driver for the development of new digital business models can offer added value for the energy industry and thus contribute to a successful energy transition. In view of the special importance of prosumers, a practical solution based on blockchain technology is to be found which will enable private producers to participate economically in electricity trading despite low output.

2 Decentralised Energy Supply

The increasing decentralisation in the wake of the energy transition is reflected not only in the emergence of new market players and roles, but also in far-reaching changes in the generation and grid infrastructures (Reck, 2011). The result is a wide range of challenges, but also opportunities for energy system transformation. The transformation of the energy system brings with it a complexity that can only be mastered with a high degree of automation. In particular, the cost of decentralised power generation and flexibility systems is increasing and requires a corresponding control and IT solution. The inclusion of a large number of individual systems in the low-voltage grid also requires an efficient communication and trading system between market participants (Verband der Elektrotechnik, Elektronik und Informationstechnik 2007, p. 91f.). This makes it necessary to collect and exchange large amounts of data. In addition to these aspects, there are also high data security requirements associated with the exchange and billing of electricity.

2.1 Introduction

With the rapid spread of technological developments in renewable energies and global challenges (climate change, migration, financial crises), decentralised production and storage options are gaining in importance. In the course of the energy revolution, the structure of the energy supply systems is changing from conventional, centralised large power plants to a structure with numerous small decentralised generation plants (DEA) and in the long term may mean an end to the monopolies by the large energy supply companies (Radtke and Kersting 2018, p. V-VIII). This will lead to a significant change in the existing energy supply system in Germany (Kühne and Weber 2018, p. 3). While in recent years decentralised energy supply has only been of interest for special applications and has seldom covered costs, considerable restructuring of the energy supply has led to a much greater interest in all forms of decentralised energy technology. The strongly changed political framework conditions, such as the nuclear consensus, the minimum feed-in tariffs for electricity generated from renewable sources or CHP, and the climate protection targets also contribute to this (cf. Beckhaus 2002, p. 1). Not only from the point of view of the end consumer, but also from an economic perspective, decentralised energy supply is to be supported as the result of a recent study by Prognos AG, the Friedrich Alexander University Erlangen-Nuremberg and the Nuremberg Energy Campus (Prognos 2016, p. 74). The rapid development of decentralised regional energy supply concepts in Germany is also mobilising regional economic structures. In order to provide the necessary energy policy impulses, attention must be paid to the balance between centrality and decentralisation in the energy supply, which will be significantly shaped by future cost structures. The following applies: the more cost-effective the components and systems for using renewable energies, the more decentralised their use will be (ForschungsVerbund Erneuerbare Energien 2010, p. 34). A study by the Reiner Lemoine Institute commissioned by the Haleakala Foundation, the Bundesverband mittelständische Wirtschaft (BVMW) and the "100 percent renewable foundation" came to a similar conclusion. Within the scope of this study, the more centralised and decentralised expansion paths for a power supply from renewable energy sources were compared in terms of their costs. As a result, from the point of view of the energy industry, the expansion of renewable energies by means of decentralised structures was to be preferred. So that all parts of Germany can benefit from the value-added effects of the energy system transformation, "a decentralised and consumption-oriented expansion of PV and wind energy onshore in the energy market should be more strongly encouraged (...)" (Reiner Lemoine Institute 2013, p. 60). An analysis of the "Development prospects of decentralised energy technologies in the electricity and heat market" also showed that the expansion of decentralised energy technologies is an important contribution to the sustainable development of the energy system. According to Welsch, in addition to broad energy efficiency activities, the long-term expansion of renewable energies and the comprehensive use of CHP potentials are particularly important prerequisites for meeting the far-reaching goals of a low-risk, economically compatible development of energy supply that protects the climate and resources (Ingenieurbüro Welsch **no year**, p. 4).

2.2   Definition and Characteristics

In the field of energy supply, different definitions can be found, depending on the perspective from which the topic is viewed. According to § Section 3 (11) EnWG, a decentralised generation plant is defined as "*a generation plant connected to the distribution network and close to consumption and load*". The Association of Electrical Engineering, Electronics, Information Technology e. V. (VDE) defines decentralised supply "*as a local, near-consumption form of supply that supplements and, if necessary, replaces the existing central supply*" (Verband der Elektrotechnik, Elektronik und Informationstechnik 2007, p. 11). In contrast to the centralised energy supply, the decentralised energy supply is characterised by the fact that the electrical energy is produced by many small DEAs and consumed directly at the place of generation. □

2.3   Benefits for the End User

Consumers achieve the greatest benefit through greater **independence**. Changes and price fluctuations affect decentralised consumers on the electricity market far less than those that are dependent on the central supplier. Therefore, in the wake of the energy revolution, more and more end consumers are deciding to invest in independent power supply and generate their own electricity. A glance at the development of the electricity price makes it clear that the consumption of energy generated by the company itself is now more worthwhile than ever. Whereas the average electricity price in 2008 was 21.65 ct/kWh, the consumer had to pay 29.44 ct/kWh in 2018, which corresponds to an increase of around 36 % in just ten years. While owners of PV systems used to benefit primarily from the statutory feed-in tariff, today it is much more worthwhile to consume the electricity generated yourself. The following applies: the higher the general electricity price, the more attractive it becomes to consume your own electricity, e.g. using a PV system (Aurbach, 2018). Irrespective of price fluctuations, it is also worth investing in your own energy sources, as this enables you to save money in the long term by producing and consuming your own electricity (Beegy, 2015). The decentralised feeding of electrical energy into the low or medium-voltage grid takes place. Operators of DEA receive a fee from the VNB in whose network they feed in. This corresponds to the **grid fees (vNE) avoided** by the respective feed-in compared to the upstream grid and transformer levels. The fees for avoided grid use were first introduced in 1999 in the Associations Agreement II and finally adopted in 2005 in § 18 StromNEV. When the legislator introduced the vNEV, it was assumed that the grid operator would incur lower expenses due to decentralised feed-in than for feed-in from large power plants. This was based on the assumption that the decentralised energy fed into the grid would not first have to be fed into high-voltage lines from large power plants and then transformed into low and medium-voltage grids until the end customer takes delivery of the electricity. In practice, however, it has been shown that decentralised feed-in from volatile energy sources such as the sun and wind does not save on grid expansion, but rather makes grid expansion necessary for connection in low load areas (Bundesnetzagentur, no year). Operators of systems that are remunerated within the framework of § 19 Paragraph 1 EEG 2017 do not receive this fee. Instead, the vNE is credited as income to the upstream TSOs and relieves the EEG levy. The NEMoG, which came into force in 2017, regulates the gradual abolition of the VNE. According to this, no more VNE will be paid for new plants with volatile electricity generation (wind and PV) once they are commissioned in 2018. In the case of existing plants (i.e. commissioned before 2018) with volatile generation, the VNE will be reduced by one third for 2018 and by two thirds for 2019. As of 2020, no further compensation will be paid for DEA with volatile generation. For new plants with non-volatile electricity generation, the vNE will only be abolished for commissioning from 2023 (University of Duisburg-Essen 2018, p. 15f.).

2.4 Challenges for Energy Suppliers and Distribution System Operators

Since the Incentive Regulation Ordinance (ARegV) came into force in 2007 and was first applied in 2009, the energy industry environment for grid operators has changed significantly. This is due in particular to the expansion of systems for generating electricity from renewables, which are often fed in at random and are largely connected to the distribution grids. In addition to transporting energy to consumers, distribution grids are also increasingly taking up decentralized power generation and distributing it to upstream grids. This change requires the regional and municipal distribution grids to be adapted to intelligent power grids in which producers, consumers, storage facilities and grid resources are interconnected (Bundesministerium für Bildung und Forschung, no year). The VNB are therefore faced with the task of **expanding and converting their networks**. In order to promote and encourage the associated investments for the expansion and conversion of the networks by the VNB, a modernised regulatory framework has been created with the ARegV. In order to ensure technological neutrality, efficiency incentives are increasingly being created. Network operators will thus benefit if they make the necessary investments in energy system transformation in an efficient manner (BMWi no year, p. 1). Since a decentralised system has advantages over a centralised system in many areas of **supply security**, it is all the more important to take into account the associated requirements on the distribution networks. While in the past distribution networks were designed for top-down power transmission, they are now faced with **increased control requirements** as decentralisation increases. Since this grid level has not yet been designed for this purpose, investments that go beyond the conventional expansion of the distribution grid will be necessary in the future to implement this coordination task. These include in particular the comprehensive ICTs between network components, generators, storage facilities and end consumers (Bauknecht et al. 2015, p. 18).

2.5 Development Goals and Future Prospects

The energy transition, which is expected to change large regions of the world in the coming decades, is marked by the abandonment of fossil resources. Fossil energy will mainly be replaced by sustainable wind and solar energy. While today's electricity supply comes mainly from large nuclear or fossil thermal power plants and hydroelectric power plants and has a high degree of supply security without fluctuations, **future electricity generation** will be characterised by regenerative sources with comparatively large fluctuations. Since generation using wind and solar energy is subject to strong fluctuations, electricity storage facilities or power plants are required to compensate. In order to minimise the need for storage or compensation generation, the planning of a generation mix and the spatial networking of different regenerative sources will gain in importance in the future. Since the renewable energy supply is characterized by a stronger performance orientation compared to the conventional energy supply, the acceptance of the population represents an important development goal. So far, renewable energy sources have been promoted and must be integrated into the grids. In the future, holistic support will be necessary. The costs of grid expansion for the integration of renewable energy sources should be taken into account in order to favour efficient plants with high full load hours and lower integration capacities. The need for short- and long-term storage capacities should also be taken into account in the integrated planning. In addition, incentives should be created for flexible consumers, such as charging strategies for electric vehicles, in order to achieve a high degree of utilisation of renewables even in times of high generation surpluses. In addition, electricity and heating networks are to be jointly optimised in order to feed surplus electrical energy into efficient end-use and thus reduce cost-intensive storage technologies (Brauner 2016, p. 32f.). While today's energy supply in Europe is still predominantly centralized and a few large power plants in the hands of a few companies supply energy to consumers via the transmission grids, the future energy management system will be more decentralized and medium-sized. This will lead to far-reaching changes, especially in the generation and grid infrastructures (Reck, 2011). The dynamic technical development is particularly evident in the concepts of energy and heat supply. The efficiency, effectiveness and feasibility of a decentralised energy supply are the result of the individual technologies and their implementation. With the further development of each individual area and component, the technical and economic attractiveness and feasibility for private households also increases. In the future, decentralised energy supply must follow the ideal of self-sufficient, independent security of supply. In combination with the increased use of renewable resources, this approach promises to solve the problems of outgoing fossil resources and environmental pollution (Heizung-Fachberater, 2018). From the point of view of classical energy supply, however, the decentralised approach also represents a system upheaval that calls into question the current business model. End customers are increasingly generating their own electricity and thus purchasing less electricity (Brauner 2016, p. 24f.). In addition to the respective taxes and allocations, the price for energy supply today comprises at least one energy-dependent price for the energy supplied and the use of the grid, as well as a basic price for the provision of the necessary recording equipment and for the associated billing. The political guidelines require that no NNE be paid for electricity generated and used by the company itself, although the costs for the grid continue to be incurred. In addition, neither concession fees nor value-added tax or electricity tax are levied on this electricity. The integration of DEA into grid regulation and the electricity market therefore requires the further development of existing **tariff models** so that plant operators can support grid operation economically. In order to better integrate DEA into grid operation and the electricity market, cost-effective solutions for generation profile measurement and billing need to be developed. The integration of a large number of individual systems into the low-voltage grid also requires an efficient communication and trading system between market participants (Verband der Elektrotechnik, Elektronik und Informationstechnik 2007, p. 91f.). The end users with their own generation have intelligent meters (smart meters) in order to have the possibility of remote reading and setting of variable reference and feedback tariffs. The recovered electrical energy will either be remunerated as subsidised green electricity, freely marketed by energy service providers or in future also directly marketed via fully automatic, networked decentralised billing systems between participants of decentralised balance groups (**direct electricity market**). In the case of high in-house generation rates, it is likely that the prosumers will have to pay an output price determined on the basis of the connected load instead of an output price determined on the basis of the energy transport. This is based on the assumption that the network costs will remain high and that the prosumers will not be able to justify a self-sufficient installation economically (Brauner 2016, p. 29ff.). The plants of decentralised electricity producers cannot compete with large power plants in terms of size, which is why they are currently unable to participate economically in electricity trading. However, this is not only due to the size of the production sites, but also to the fluctuations that occur in PV systems or wind turbines due to weather conditions. In order to become competitive, various decentralised producers have to join forces and form a virtual power plant. For a nationwide supply, the resulting virtual power plants should in turn network with each other in order to be able to compensate for bottlenecks. This is made possible by an EnMS, with which the various generation systems can be bundled. Additional operating and monitoring functions enable the anticipation of supply bottlenecks and the integration of possible energy storage systems. Additional energy storage systems make it possible to store excess energy and compensate for peak loads. However, complete independence from the central power grid is currently only possible with an isolated solution, which is not very practical in Germany and is also technically difficult to implement. A good alternative is the intelligent use of the public power grid by so-called **energy communities**. If the self-produced electricity is not sufficient, the remaining electricity can be obtained from the energy community. The electricity generated from renewable energy sources in a sustainable and regional way can be used to provide an independent power supply compared to conventional power plants (Gridx, 2017).

3 Digitalization of the Energy System Transition - Energy System Transition 2.0

The digitalization of the energy system transformation plays a central role for the collection and exchange of large amounts of data as well as with regard to the high demands on data security associated with the exchange of electricity and the associated billing of electricity. Intelligent metering systems, such as the smart meter and the SMGW required for communication, can be used to collect, transmit and process data on power generation and consumption in real time (Maier 2018, p. 3). According to § 29 Para. 3 MsbG, the Smart Meters are to be installed step-by-step and area-wide for all consumers by 2032. Larger consumers and generation plants are taking on a pioneering role in the use of modern measurement and control technology. Smaller electricity consumers will follow later. The SMGW is regarded as the key technology for digitalizing the energy revolution. It communicates with various components and participating market players (BSI, no year) for the transmission of consumption data as well as for its administration. The SMGW guarantees data protection and data security at the highest level and lays the foundation for an intelligent and secure network.

3.1 Introduction

The first phase of the energy system transformation has considerably accelerated the conversion of electricity supply to renewable energies in Germany. Whereas in the past electricity flowed only in one direction and information about current flows was very limited, the decentralised power supply system of the future will be characterised by bidirectional information and current flows. This is increasingly leading to a mixture of market roles. More and more passive (electricity consumers) are developing into **prosumers** who, for example, actively participate in the design of the electricity supply system with PV systems on private homes. This change from consumer to prosumer is a further step on the way to emancipating energy customers from the established energy supply companies, which will result in changes in the supply of energy and accompanying services (Servatius et al. 2012, p. 288f.). Increasing decentralisation, changing load flows in the grids, an increasingly volatile energy supply, a large number of new market players and roles as well as integrated services place high demands on the control of electricity and gas grids and the guarantee of supply security throughout the entire energy system. Energy suppliers are currently facing major challenges such as the energy transition, increasing competition and declining yields in the electricity market. In order to meet these challenges, digitalization, i.e. the transformation of analogue information into a digital form, is necessary. This is particularly useful in the energy sector due to the large amount of data (performance data from power plants, consumption data, temperature data, weather data, etc.). In particular, digitalization can be an "enabler" for flexibilization and help to better integrate volatile renewable energy sources (Zimmermann and Hoppe 2018, p. 11). In this context, the digitalization of the energy system transformation plays a central role, for example in increasing the efficiency of business and work processes, developing new business fields, customer loyalty, reducing costs and maintaining grid stability (Roth 2018, p. 4). For the Fraunhofer Institute for Experimental Software Engineering (IESE), "*a successful energy transition without comprehensive digitalization is inconceivable*" (Neugebauer 2018, p. 347). The digitalization of the energy system transformation requires numerous IT products, complex control and distribution systems or integrated communication and monitoring applications (Kulturwissenschaften, no year). However, all this is now leading to a further transformation of the energy world - an "Energy Transition 2.0" - with which a new, digital level is being introduced.

3.2 Smart-Energy

The digitalization of the energy industry results in particular from the technological framework conditions of the energy system transformation. The paradigm shift from central power supply to bidirectional or multidirectional power distribution requires a new infrastructure that can only take all possible state scenarios into account through the massive use of ICT. In view of the current political discussion and decision-making situation regarding the phasing out of nuclear power, the establishment and expansion of renewable energy generation, the provision of adequate power grids and the use of individual energy saving potentials, no one doubts the necessity of implementing intelligent energy (smart energy) (Aichele 2011, p. VII). Smart-Energy is a collective term for so-called intelligent technologies along the entire value chain (generation, networks, trade, distribution, consumption) in the energy industry. Smart-Energy consists of Smart-Meter (intelligent meter), Smart-Grid (intelligent power grid), Smart-Market (intelligent market), Smart-Home (intelligent house), and Smart-City (intelligent city) (ITWissen, 2018). Fig. 3.1 shows the respective components of smart energy, which are explained in more detail below.

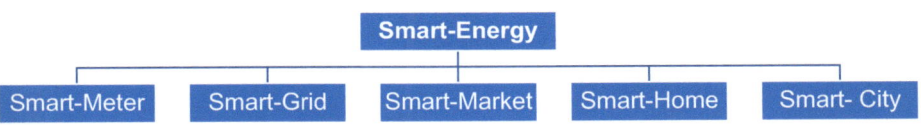

Fig. 3.1: Components of Smart-Energy. Source: Own presentation based on Holstenkamp and Radtke (2018): p. 700ff.

Smart Meter

The Smart Meter is a digital electricity meter that has been used in companies since the 1990s and has been increasingly found in private households since the beginning of this decade (Meta-Level, no year). The installation of these smart meters should enable secure and standardized communication between different network actors (Zimmermann and Hoppe 2018, p. 11). Depending on the equipment, they can provide end consumers, network operators and producers with the necessary consumption information, be used to transmit network status data, support secure and reliable control measures and serve as a kind of communication platform in the intelligent energy network. By obtaining precise information on the consumption behaviour of the final consumer, smart meters can serve as a tool for greater energy efficiency. They make energy consumption visible, ensure flexible use of energy and thus motivate people to save energy. The development of smart meters is now well advanced. In order to make the best possible use of the technology, rules and obligations are required for all those involved. These rules and obligations were laid down in 2016 by the legislator in the Act on the Digitalization of the Energy Transition (GDEW) and the Measuring Point Operating Act (MsbG) contained therein (BMWi, 2016). The new regulations create a balanced relationship between the costs and benefits of installing and operating intelligent measuring systems (iMSys). For this purpose, there is a clear cost regulation with price caps which ensure that the costs do not exceed the expected benefits (BMWi, 2015a). The MsbG forms the core of GDEW and regulates how and when the existing analogue electricity meters will be replaced by iMSys or modern measuring equipment (Eon, no year). The focus is therefore on the introduction of iMSys, which serve as a secure communication platform to make the power supply system suitable for energy applications (BMWi, 2015a). According to § 29 Abs. 3 MsbG the digital electricity meters are to be installed by 2032 area-wide and gradually with all consumers. Larger consumers and generation plants will assume a pioneering role in the use of modern measurement and control technology. Smaller electricity consumers will follow later. The experience gained in the high-consumption groups can then be used to gain experience in the household sector as well. GDEW also guarantees the **minimum technical requirements** for **data protection** and **data security** (SWD, no year). On behalf of the BMWi, these were developed by the Federal Office for Information Security (BSI) together with industry representatives in close cooperation with the Federal Commissioner for Data Protection and Freedom of Information, the BNA and the Physikalisch-Technische-Bundesanstalt. The documents can be viewed on the BSI website and contain a technical standard for **"privacy-by-design"**. The law also contains detailed regulations on who may access which data and when. On the one hand, this ensures a high level of data protection; on the other hand, all data that is absolutely necessary for energy supply can be used by the respective authorised actors (BMWi,

2015a). In an iMSys, the Smart Meter Gateway (**SMGW**) with integrated security module forms the communication unit. The measured data is received from meters, stored and prepared for market players. The SMGW is regarded as the **key technology** for digitalizing the energy revolution and communicates with various components and market players involved (BSI, no year) for the transmission of consumption data as well as for its administration. The SMGW guarantees data protection and data security at the highest level and lays the foundation for an intelligent and secure network. With the first certified SMGW, the BMWi and the BSI show that digitalizing is successful even with high data protection and information security requirements, even if the legally prescribed roll-out of the SMGWs does not begin until three devices from different manufacturers have been certified by the BSI (Windmesse, 2018).

Smart-Grid

As a result of the increase in decentralised, mostly regenerative generation plants, it will become much more complex in future to adapt the necessary base load, standard load and peak load generators to the consumption load profiles. The prediction of the exact generation volume becomes less accurate due to the dependence on climatic conditions and numerous decentralised as well as individually controlled generation plants. Bidirectional communication with consumers and the control of consumption behaviour through short-term adjustments to electricity prices are intended to align the generation and consumption curves (Aichele 2011, p. 64f.). So-called Smart Grids (SG) are used to ensure energy supply on the basis of efficient and reliable system operation. An SG leads to a better utilisation of the conventional network infrastructure, which dampens its expansion requirements or improves network stability at the same capacity utilisation. With regard to distribution networks, this term refers to the increasingly improved possibility of tracing system states in the network and intervening locally. In addition to ensuring the supply of consumers from both local anc supra-regional sources, this means an increased possibility of receiving regionally generated electricity and passing it on to higher voltage levels without loss of network security. In smart grids, for example, capacities can be increased or flow directions can be changed over individual line sections. According to the BNA, these structures should also create the bas s for future market opportunities also for small grid users without negatively affecting grid security (Bundesnetzagentur 2011, p. 11f.).

Smart-Market

With the necessary integration of the substantially growing share of fluctuating, stochastic energy sources in total electricity generation, solutions for their efficient market integration are more in demand than ever before. Due to its proven ability to comprehensively integrate volatile renewables into the supply landscape, a functioning intelligent energy market (smart market) is logically assigned a key energy policy function that contributes to the success of the energy transition (Aichele and Doleski 2014, p. 36).

Smart-Home

Smart-Home (intelligent house) is generally understood to mean improving the quality of life and living, many ways of saving energy in the house or apartment and increasing safety. The networking of home technology and household appliances (heating, refrigerator, stove, washing machine, lamps, etc.) as well as the networking of consumer electronics (television, radio, etc.) also fall under the umbrella term Smart Home. One example is intelligent heating control, which in the simplest case can be the replacement of the manually operated valve on the radiator. The intelligence lies in a time control, in recognising whether a window has been opened or in any other freely programmable profile in order to intelligently regulate the heat supply to the radiator. In an intelligent building, the entire heating circuit or the air conditioning supply is therefore optimally regulated in terms of energy technology, resulting in noticeable cost savings after just one year and the investment being amortized after just a few years (Aichele and Doleski 2014, p. 511f.). Applications close to the market around the Smart-Home can make a considerable contribution to increasing consumption transparency, energy efficiency and grid stability by smoothing out consumption peaks in subsequent years. By using energy management measures adapted to the needs of household customers, the end customer can also be integrated into the smart market. Energy management primarily comprises largely automated household installations for appliance control, which enable dynamic adaptation of the respective energy use to the actual supply situation with the aid of price signals (Aichele and Doleski 2014, p. 32). The Smart-Home will fully unfold when the environment adapts to the needs of its residents or tries to anticipate them. New business models, technologies and application scenarios shape the dynamics of the market (Gründiger, no year). So far, however, the selection and sale of intelligent energy technologies for the household has been limited. According to a study by the Federal Statistical Office, the proportion of households equipped with smart home applications in relation to the total number of private households in Germany in the respective segments is in the single-digit percentage range (Statista, 2017).

Smart-City

The term "Smart-City" (intelligent city) means *"...as a rule, all concepts of modernising cities and making them a better place to live with the help of the possibilities offered by new technological developments and information and communication technologies with regard to ecology, social coexistence, political participation, etc."* (Siepermann, no year). The focus of a smart city is on the efficient use of energy resources in urban settlements. Transparent processes, intelligent controls and integrated information flows enable energy-optimised operation of the entire urban infrastructure and the connected technical systems. Numerous instruments of the smart market contribute directly to the gradual development of conventional cities into networked, energy-optimized and sustainable spaces of urban human life. The further development of urban structures will depend decisively on the technical possibilities of ICT and the ability of these technologies to control supply and disposal networks efficiently (Aichele and Doleski 2014, p. 31). The "Smart-City" component is very comprehensive and goes far beyond support for local energy system transformation. As far as local energy system transformation is concerned, Smart-City can, for example, support decentralised energy generation and network local producers with each other (Kühl, 2018). The city of Duisburg is setting a good example as part of the "Smart City Duisburg" concept. The concept for developing the "Digital Duisburg Master Plan" was developed in 2017 on the initiative of the Lord Mayor of the City of Duisburg and Duisburger Versorgungs- und Verkehrsgesellschaft (DVV). It uses ICT to intelligently link municipal infrastructures such as energy, buildings, traffic, water and wastewater. The information flow between the different infrastructures is analyzed and translated into services for citizens and organizations. The transformation into a Smart-City should increase the economic efficiency and the quality of life in the city. This will be achieved by expanding a free W-LAN network across cities, optimizing local public transport and introducing e-government to simplify administrative procedures (Smartcityduisburg, 2017).

3.3 The Challenges of Digitalization

The phasing out of nuclear energy and the associated expansion of renewables mean that German energy suppliers face a number of challenges, particularly in their core areas of generation, trade/distribution and grid business. For example, additional digital processes cause **high computing costs** and **energy consumption**. In addition, the **market roles** and **responsibilities** of new players in electricity trading are still **unclear** (Maier 2018, p. 3). With regard to the widespread use of DEA in the distribution network, a **lack of communication facilities** for controlling and monitoring the distribution networks poses a major challenge (Verband der Elektrotechnik, Elektronik und Informationstechnik 2007, p. 47). Due to the increased need of customers for individual solutions and above all for technical developments such as highly scalable, intelligent IT platforms, **new competitors** with high IT expertise are forcing their way into the existing market (Wienhold, 2016). In order to advance digitalization, existing analogue electricity meters must be replaced by iMSys or modern measuring equipment. This requires specifications on **minimum technical requirements** in the form of generally binding protection profiles and technical guidelines. The **regulation of the operation of measuring points** and the **equipping of measuring points with modern**

measuring equipment and iMSys will also be necessary in order to set the framework for cost-efficient, energy-efficient and consumer-friendly future measuring point operation. **Data protection** is particularly important. According to § 21g EnWG a regulation of the permissible data communication is necessary to guarantee **data protection and data security** in modern energy networks. There are concerns that an increasing technological dependency due to digitalization will also increasingly dominate the private sphere and create a feeling of external control. Current tendencies to control the population through the expansion of digitalization, such as in China, reinforce these fears. In addition, hacker attacks could also occur in the energy context (Zimmermann and Hoppe 2018, p. 13). Unauthorized external access to objects in the digital world can lead to personal and personal data being tapped and, in the worst case scenario, to incorrect control being provoked. Unauthorized access to generation and distribution systems in particular can cause considerable economic damage through reprogramming or shutdown. **Security aspects** and **trustworthiness** are therefore decisive factors in the acceptance of new solutions (Aichele and Doleski 2014, p. 504f.). Vohrer, too, sees the *"(...) conflict of goals between data protection and the data hunger of an intelligent energy system (...)"* (Müller, 2018) as a central challenge. The question of the economic or political use of data has not yet been sufficiently clarified either. If the newly collected data converge at a few companies, there is a danger that these companies will gain increasing market power (Zimmermann and Hoppe 2018, p. 13). Since modern, digitally controlled energy systems with networked infrastructures are also vulnerable, the energy system must be **resilient** and **flexible** (Renn, no year). Although flexible networking can be seen on the one hand as increasing resilience, it also requires an increased use of ICT. This involves the handling of large amounts of data and at the same time the **risk of external attacks** on digital energy infrastructures, which endanger energy security (Wiegand 2017, p. 13). Due to the change in the electrical energy supply, the currently provided generation **flexibility** of easily controllable large thermal power plants will no longer be available to a sufficient extent in the future. This lack of control potential must therefore be compensated for by increased flexibility of the supplied consumers and decentralised feeders as well as by energy storage systems (Zdrallek et al. 2016, p. 9). Increasing decentralisation is also reflected in the emergence of new **market players** and roles, such as new suppliers, decentralised producers or metering service providers, which means that greater coordination is required to **guarantee security of supply** throughout the entire energy system (Servatius et al. 2012, p. 288f.). According to a study by the Hans Böckler Foundation on the subject of "Digitalization in the energy industry", developments will have far-reaching consequences for work and employment in addition to technical challenges. In addition to the discontinuation of individual activities, the increasingly digital, flexible and networked provision of services in the energy industry will also lead to further changes in work organisation, work content and qualification requirements. Accordingly, the energy industry faces numerous challenges. The digitalization of business and work processes to increase efficiency in technical and administrative processes, new digital models for customer care, the implementation of digital technologies for the operation of new energy networks (smart meters, smart grids) and the development of innovative services are of outstanding importance. These include, for example, the improvement of energy efficiency, intelligent networking and control of building services (smart home), new price models and the use of big data (Roth 2018, p. 4-7).

3.4 Development Goals and Future Prospects

In principle, digitalization can make the energy system more efficient. This is also necessary in order to make the energy revolution a complete success (Müller, 2018). However, as a result of the study of various publications, Vohrer finds that the energy industry lags far behind other economic sectors in terms of digitalization (Umwelt-Energie, 2018). The study by the Hans Böckler Foundation on the subject of "Digitalization in the energy industry" also makes it clear that the energy industry is only at the beginning of a digital transformation. Approaches are discernible, but how the industry develops and what effects digitalization will have in the future depend on a number of factors, such as the selection and objectives with which digital technology is used, the corporate strategy and the positioning of the company's representation of interests (Roth 2018, p. 6). What is certain, however, is that as the number of decentralised feed-in and supply systems within existing networks increases, the coordination requirements and the amount of data to be controlled will increase significantly. Decentralised generation will further increase control complexity and the load on the grid infrastructure in the future. At the same time, the increasing use of renewable energies will increasingly lead to capacity fluctuations in the electricity grid. Energy suppliers are thus faced with the challenge of ensuring a continuous, largely stable balance between electricity supply and demand at all times despite increasing fluctuations (Roth 2018, p. 7). As a result of rising energy costs, more and more house and building owners are opting to generate their own electricity (Aichele and Doleski 2014, p. 511f.). As a result, many grid customers will develop from pure consumers to **prosumers** in the future. They thus form the starting point and (de)central element of the future energy system. Prosumers not only obtain energy from the grid, but also provide energy for the grid or market it directly in certain periods of time. As a result, the load behaviour of grid customers changes significantly, making active grid customers a central pillar of future energy supply. In the long term, the respective grid structures will also change. The future energy supply will increasingly be provided by networked local supply units, some of whose energy requirements will be secured by these themselves. In the event of energy demand or surplus, a balance is created by networking individual supply units. The size of the respective unit, the degree of self-sufficiency and its characteristics (households, commercial, trading, service and industrial enterprises) can vary regionally. In principle, the sector coupling in the energy supply system mentioned at the beginning must also be taken into account, in which a supply unit is characterised not only by supplying conventional electrical consumers but also, for example, by providing heat or mobility. In order to adapt the networks to the resulting requirements, ICT and **innovative technologies** and planning approaches are increasingly needed (Zdrallek et al. 2016, p. 9). From the point of view of the **end customers** (prosumers) involved, a number of **requirements** will have to be met in the future for the technical and energy-economical development of decentralised energy supply systems. Thus, the surplus energy or the purchased balancing energy is to be exchanged as directly as possible or in real time within the same grid level in order to guarantee a minimum grid load and expansion requirement for the higher voltage levels. In addition, suitable **automatic billing systems** for prosumers should exist in both the private and the public grid. Since both energy and related financial transactions take place automatically, operation and maintenance must be carried out on a fiduciary basis. At this point, Brauner suggests that this should be done by the NSOs, as these are not allowed to trade electricity as a result of deregulation and therefore no conflicts of interest can arise. Supervision of this must be carried out by a neutral body, e.g. the energy regulatory authority. Intermediaries would, however, reduce the degree of an automated and self-regulating market, which is why alternatives must be sought. By parameterising the automatic billing systems accordingly, each customer should continue to be able to freely choose the supplier. In the event of a malfunction of the automatic billing system, it must be possible to return to **historical counting and measurement values**, which is why Brauner says smart meters with memory are required. The exchange of energy between parties within a building can be described as a private-sector balancing group. If, on the other hand, a building participates in an energy exchange via public networks, it must become a member of a balancing group. Since many **small balance groups** can form in the future, an extensive **automation** of small groups is necessary. Since operation within the distribution network and electricity exchange within and outside the network, as well as the associated energy billing, are to take place fully automatically by networking smart meters with decentralized energy management and billing systems, high demands will be placed on **data security** in the future (Brauner 2016, p. 30f.). These aspects make it clear that the digitalization of the energy system transformation is accompanied by increasing complexity and decentralization. GDEW, which came into force in 2016, is intended to create the basis for the digitalization of the energy system transformation. In order to monitor the implementation status of the law and the progress of the digitalization process, the BMWi set up a monitoring system at the end of 2017 with the service contract *"Digitalization of the*

Energy Transition - Barometer and Top Issues". The Barometer, which is published annually, will also examine ways of improving and accelerating existing procedures and processes. The first annual barometer was published on 30 January 2019 and provides an overview of the technological solutions in GDEW's areas of application. In addition to modern measuring equipment, cross-divisional metering including sub-metering, control of generation plants and consumption equipment, smart mobility and smart home, **blockchain technology** was also listed as a technological solution. Blockchain technology offers novel solutions and could prove revolutionary due to its ability to replace intermediaries with programmed codes and thus reduce transaction costs (Zimmermann and Hoppe 2018, p. 13). As a result of a BDEW study on *"Blockchain in the energy industry"*, blockchains *"(...) have the potential to optimise energy industry processes in almost all stages of the value chain and at the same time cope with the increasing complexity in the increasingly decentralised energy system"* (Bundesverband der Energie- und Wasserwirtschaft 2017, p. 6). In particular, the consistent establishment of market mechanisms in the course of the advancing energy system transformation means that very small amounts of energy or flexibilities are traded and the contractual partners (mostly prosumers) are not able to conclude bilateral framework agreements first. The necessary contractual agreements to secure large quantities of short relationships between the many players require machine-concluded contracts that are based on the framework agreements concluded by humans. Here, too, there is a need for research in order to make legally secured agreements in real time between machines, which must then be translated into system actions, monitored during fulfillment, comprehensibly documented and correctly invoiced. According to Neugebauer, the blockchain approach is also propagated in this context (Neugebauer 2018, p. 361f.). Blockchain technology opens up new business opportunities in the energy sector, which are dealt with in the sixth chapter. Blockchain technology is seen as a hope for a small-scale trade between prosumers, with which the grid should regulate itself and harmonize production and consumption (Kühl, 2018). In order to understand the purpose of blockchain technology, the basics of this technology are described below.

4 Basic Principles of Blockchain Technology

The digitalization of the energy industry is constantly evolving. Blockchain technology is currently the focus of attention as a new driver of this rapid development. The functionality of the blockchain is the decentralized storage and encryption of transaction data in a long chain of data blocks. Different types of blockchains can be used, which can be classified in terms of data access and network usage. In addition, there are different consensus mechanisms, the use of which depends strongly on the respective use case and the trust in the blockchain network. Blockchain technology has been further developed in recent years, resulting in new use cases. Smart-Contracts in particular contribute to the automatic conclusion of contracts and to making business processes more efficient and simpler.

4.1 Introduction

Reliability and trust are the decisive core elements for the digitization of business processes along the entire value chain. Traditionally, databases and process management have always followed a centralized approach based on an appointed authority with a centralized process synchronization. However, this centralization is associated with a number of risks, such as performance bottlenecks, reliability, authenticity, and internal and external attacks on integrity. Advances in the development of ICT and the Internet over the past decades have enabled a number of innovations that have a major impact on almost every sector of society. Examples include social media innovations, crowdfunding platforms and the digital currency Bitcoin, which has attracted public attention in various contexts in recent years. Especially the aspect of carrying out transactions through a distributed computer network without the influence of a central institution caused a sensation. According to experts, the underlying blockchain technology has the potential to change many areas of society and represent the next disruptive innovation due to its characteristics (Schlatt et al. 2016, p. 5). Blockchain technology thus has great relevance for the digitalization of services and processes and could fundamentally change numerous industries in the coming years (Neugebauer 2018, p. 311). It is not uncommon to be attributed the potential to change the world to the same extent as the Internet has done since 1990. Blockchain applications rely on distributed networks, cryptography and game theory and promise not only transparency and manipulation security but also significant cost reduction potential by replacing intermediaries with software solutions.

Definition

So far, no uniform definition of the block chain has been established. However, there are numerous attempts by different authors to define the blockchain. Walport defines a blockchain as a kind of database in which entries are grouped into blocks and linked in chronological order by a cryptographic signature. In technical jargon, blockchain technology is referred to as Distributed Ledger Technology (DLT). Walport differentiates the blockchain from distributed ledgers, in which records are continuously sorted and stored instead of in blocks (cf. Walport 2015, p. 17). However, since the terms blockchain technology and DLT are often used synonymously in science and practice, no distinction is made in this paper either (BaFin, 2017). In simple terms, the blockchain can be understood as a jointly written digital account that documents and verifies transactions (Zimmermann and Hoppe 2018, p. 15). According to Burgwinkel (2016), in the so-called distributed approach the data is not stored in a central database, but distributed to the systems of the network participants, and cryptographic procedures are used to ensure that the integrity is given. Condos et al. (2016) define a blockchain as an electronic register for digital data records, events or transactions that are managed by the participants of a distributed computer network. Glaser and Bezzenberger describe the associated management systems as distributed consensus systems which, according to the authors, are based on cryptography and peer-to-peer (P2P) principles rather than on a central authority to achieve network-wide verification of the status of the system by consensus. According to the P2P principles mentioned above, network participants provide hardware resources to provide content or services of the network. In addition, there is a direct exchange between the computers (nodes). This means that there is no central authority to coordinate communication between the individual nodes (Glaser and Bezzenberger 2015, p. 2). According to Laurence, the blockchain represents a data structure that makes it possible to create a kind of digital ledger with data and to share it via a network of independent parties (Laurence 2017, p. 23). For Mitschele, the blockchain is a decentralised database that is mirrored on a large number of computers in the network and is characterised by the fact that its entries are grouped into blocks and stored. In addition, the authenticity of the database entries is ensured by a consensus mechanism used by all computers (Mitschele, no year). Drescher (2017) provides a solid definition that provides a comprehensible overview: *"The blockchain is a purely distributed peer-to-peer system of general ledgers that uses a software component consisting of an algorithm that negotiates the information content of ordered and connected data blocks together with cryptographic and security technologies to achieve and maintain their integrity"*. The above definitions indicate that blockchain systems use cryptography. The consensus mechanism also mentioned above, by means of which the network nodes coordinate the system status, can be seen as the fundamental innovation behind blockchain systems. Furthermore, the above definitions indicate that blockchains are based on a **distributed network**. Mullender writes that distributed systems cannot be exactly defined, but rather characterized by several properties (Mullender 1990, p. 32). The characterization is then performed by delimiting different network architectures.

Distributed Networks

One of the fundamental decisions in system implementation concerns the architecture, i.e. the way in which the individual components are organized and interrelated. The three essential network architectures for software systems are the centralized, the decentralized and the distributed network, which are represented graphically in Fig. 4.1.

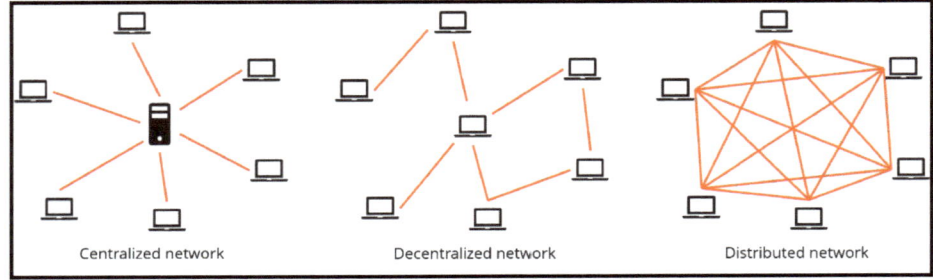

Fig. 4.1: Network structures. Source: Ethereum-base (2017).

In a **centralized** network, the components are not directly connected to each other. Instead, there is always only one direct connection to the central component. Basically, this type of network is particularly suitable for communication. A coordinating instance ensures that all participants can communicate and adhere to certain rules. It also ensures that there is a uniform view of the data. In a central system, all participants trust the central instance. If the trust with the central authority is compromised, there is a risk of the system collapsing. In a decentralized network, coordination takes place between the participants themselves. This type of network is therefore particularly suitable for interaction. In order to be able to interact together, a communication protocol is required which must ensure the integrity of the system. Integrity means that the data is correct. However, the large number of different participants in the **decentralised** network can lead to a number of potential malfunctions if, for example, two people edit a file at the same time and the system has to decide on a version. Since the participants in the decentralized network do not know each other and therefore no trust exists, the system must be protected against manipulation. There has been no satisfactory solution for these problems so far, which has led to the fact that today's Internet is predominantly centralized. Blockchain technology differs from today's Internet in that it is neither centralized nor decentralized, but instead distributed. **Distributed** networks initially consist of several independent nodes that communicate and synchronize with each other. The failure of individual nodes does not affect the functionality of other nodes. In addition, each node stores a common status of the system, so that the failure of individual nodes does not imply the loss of the system status. In blockchain systems, the data of the blockchain is stored redundantly in each node. Blockchains can thus be described in simplified terms as distributed network architecture in which all components are indirectly connected to each other without a central element. The focus here is on transactions which can be very different and can be based on crypto currencies, stocks, documents or goods, for example. The blockchain technology ensures on the one hand that fraud is impossible, so that trust between the participants is no longer necessary. On the other hand, it ensures that the data is correct and that there is a uniform truth for all participants (Breitsprecher, 2018). Blockchain technology thus ensures integrity in a decentralized network with an unknown number of subscribers who do not know each other and therefore do not trust each other.

History and Milestones of the Blockchain

The concepts on which blockchains are based are based on more than thirty years of research. Already in **1979** Ralph Merkle invented the principle of hash trees, which are also called "*Merkle trees*". In **1991**, Haber and Stornetta published a scientific article on how to time stamp and link documents. This principle is also known as *"linked timestamping"*. In **1997**, Nick Szabo published his vision of smart contracts to show how e-commerce could evolve and how contract processes on the Internet could be supported. Ten years later **(2007)**, the blockchain *"Guardtime"* developed in Estonia was put into operation as a commercial system. Only one year later **(2008)** an author published the article *"Bitcoin: A Peer-to-Peer Electronic Cash System"* under the pseudonym Satoshi Nakamoto. The article describes how the Bitcoin system works. The technology was also developed as a reaction to the global financial crisis of 2008, which severely damaged confidence in the financial sector and, in particular, caused the reputation of banks as trustworthy institutions to fluctuate. In **2013**, Vitalik Buterin founded the *"Ethereum"* project. Like Bitcoin, Ethereum is based on blockchain technology. In contrast to Bitcoin, Ethereum is not a pure crypto currency, but also a platform for decentralized applications (dApps) consisting of smart contracts (Burgwinkel 2016, p. 12). In **2015**, the Hyperleger Project was founded, which is a cross-departmental open source initiative to promote cross-industry blockchain technologies. Hosted by The Linux Foundation, this global collaborative project brings together leading companies from industries such as banking and finance, Internet of Things, supply chain, manufacturing, and IT. In the same year, Microsoft and IBM added blockchain applications to their cloud services offerings.

4.2 Functionality of a Blockchain

The five operations in Fig. 4.2 illustrate how the blockchain works. After a transaction has been **initiated** by a network node, the transaction is **passed** to the network and **distributed**.

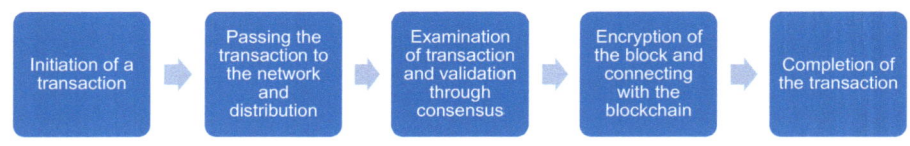

Fig. 4.2: Functionality of a blockchain. Source: Own representation based on Neugebauer (2018): p. 313.

The nodes of the network **check** the validity of the transaction and try to find a consensus. **Consensus building** determines which block is taken over as the next element in the global blockchain. In this process, the transactions are stored, which are converted into a standardized format by **hash functions** and finally compressed hierarchically into blocks. Then the checked blocks are connected with the already existing history of the blocks so that a chain is created - the so-called blockchain.

Hash Functions and Hash Values

The term **hash** is used primarily in the field of mathematics and computer science and means something like chopping something up or scattering something. Hash functions are used, for example, for **data storage**, **checksum verification** for **information integrity**, and in **cryptology**. With so-called **hash functions**, large input values can be encrypted and stored in a small target value (**hash value**). Since there are practically no two input values that generate the same target value, hash values are **collision-resistant**. This makes hash values very easy to verify. If the parameters used for encryption are known, it can be determined **very fast** whether the hash value to be checked is correct (Strasbourg, 2018). If the input is identical, the resulting hash values are also identical. If a small detail in the input is changed, the resulting hash value (output) changes completely and unpredictably. Since the blockchain is largely based on **hash references**, a comprehensive knowledge of this component is important for understanding the blockchain and how it works. Hash references are created by the combination of hash values and the storage locations of the data sets. They also check whether the data being referenced (the reference target) has not been changed since the reference was created. If the reference target is changed, the corresponding data can no longer be retrieved with the reference, which means that the hash reference is then considered damaged or invalid. However, hashing is not only used to compare information. Hashing is also used to detect changes to data that should remain unchanged. Within the blockchain, hashing is used, for example, to store data that is sensitive to changes in a tamper-proof way. Hash values can also be used to challenge computers to each other with complicated tasks. This form of using hash values is one of the most important concepts of the blockchain (Drescher 2017, p. 104-110). In the simplest case, hash references (circles marked R) are used in a chain of linked data (Fig. 4.3).

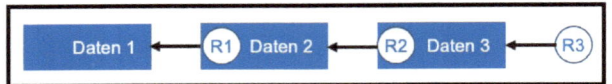

Fig. 4.3: Structure of a chain. Source: Own illustration based on Drescher (2017): p. 105.

Such a structure is useful, for example, when storing and linking data if this data is not all available at the same time. The creation of such a chain starts with the first data link "Data 1" and the creation of R1. Since this is the first link in the chain, it does not contain a hash reference. When new data is received, it is combined with the hash reference pointing to the Data 1 link. R2 refers to the newly added data and R1. R3, which refers to the "Data 3" element and R2, is generated in a similar way. Only R3 is required to access all data in the chain (in reverse order of addition). In this example, R3 forms the so-called list header, since R3 refers to the last data link added (Drescher 2017, p. 104f.). In this case, the term header must not be confused with the term header, which is used in the following Merkle tree. Such a structure is also referred to as a hash tree or, in the English-speaking world, after its discoverer, the computer scientist Ralph Merkle, as a Merkle tree. With a Merkle tree, the hash references (R1 to R4) are first generated for the individual transaction data, which are then grouped in pairs (Fig. 4.4).

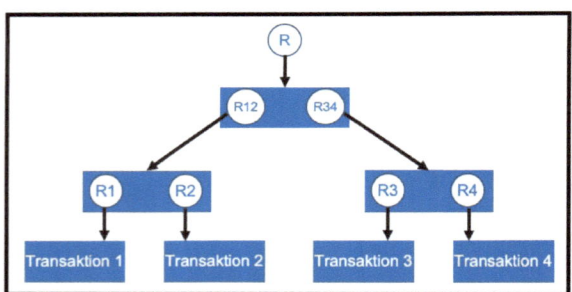

Fig. 4.4: Structure of a Merkle tree. Source: Own illustration based on Drescher (2017): p. 105.

As a result, new hash references (R12 and R34) are created. This process of combining two hash values is performed with several transactions until there is only one hash value (R) left, which simultaneously represents the so-called Merkle root of the hash tree. According to Bashir, the Merkle root represents the **header** of a block in blockchains, which comprises the hash of all transactions in the block. In a blockchain data structure, each block header usually also contains a hash reference to the header of the previous block, the degree of difficulty of the hash puzzle, the point in time at which the hash puzzle begins to solve, and a nonce that solves the hash puzzle (Drescher 2017, p. 155). The decisive advantage of the structure is that it is only possible to validate all transactions in a block simultaneously using the header, instead of checking all transactions individually (Bashir 2018, p. 19). This makes this principle particularly suitable for grouping many individual data parts if they are available at the same time and are to be retrievable via a single hash reference. If a single statement were changed, the hash tree would no longer be consistent overall. A corrupted reference in such a construct proves that some of the data was changed after the structure was created. Otherwise, it can be concluded that the entire construct has not been changed since it was created (Drescher 2017, p. 105f.). In this way, a tamper-proof concatenation can be ensured.

Validation by Consensus Building

The Blockchain is a purely distributed P2P system accessible to all. There is therefore a risk that the transaction data history may be manipulated or falsified (Drescher 2017, p. 152). In order to protect this transaction data history of the system against counterfeiting and manipulation, so-called **consensus mechanisms** are used. A consensus mechanism can be defined as the mechanism by which a blockchain network reaches a consensus. According to Duden, a consensus is described as the *"agreement of opinions"* (Duden, no year) of several parties. Since blockchains usually do not depend on a central instance, the nodes have to agree on the validity of transactions. The consensus mechanisms ensure compliance with the protocol rules and guarantee that all transactions in the block chain are processed reliably (Binance, 2018). A consensus must be reached in each block chain in order to determine which procedure is to be used to create new blocks and attach them to the existing blockchain. This is to ensure that blocks are correctly and irreversibly chained to form a blockchain. If new blocks are created to supplement the existing block chain, a consensus on the change can be reached in the entire block chain network. Since a blockchain is a distributed network, the agreement of all members of the blockchain is to be called a distributed consensus decision. The consensus mechanism used depends in particular on how trustworthy the blockchain network is. In the meantime, there are a number of different consensus models with different development statuses and individual advantages and disadvantages. The currently most important concepts are Proof of Work (POW), Proof of Stake (POS) and Proof of Authority (POA). In the following, the principle of validation by consensus building is explained using the example of the most commonly used POW. With the POW procedure, a considerable amount of computing effort is required to write or add blocks to the blockchain to ensure that the content of the blockchain data structure remains unchanged (Drescher 2017, p. 154). To add a list of transactions in the form of a block, proof must be provided that work has been done. For the POW, a cryptographic puzzle must be solved which is also known as a hash puzzle. From the block a SHA-256 hash value must be calculated, which corresponds to a certain pattern (Lang and Karlstetter, 2017). In technical jargon, this process is also known as **mining**, which has proven itself due to its robustness and security. With the POW method, the probability of finding a valid hash value for a new block depends on the computing capacity used, the so-called miner (Seidel 2019, p. 76). Due to this energy-intensive mining process, the network participants within public blockchains must be offered an economic incentive to carry out this validation. In the best-known POW application, the so-called miners receive the crypto currency Bitcoin, which is paid out to participating network participants and thus remunerates them for the validation (Strasbourg, 2018). The challenge of an unchangeable blockchain data structure is to make adding a new block (artificially) a computationally intensive task (hash puzzle). Basically, each block header must contain a **hash reference to a previous block**. There must also be a **valid root of a hash tree** that contains transaction data. In addition, the **time stamp** of the block header must be after the time stamp of the previous block header. The timestamp ensures that the blocks and transaction data are actually sorted in the order of the times of the entries. Fig. 4.5 below shows the basic structure of a hash puzzle that must be solved when entering a new block in the blockchain data structure.

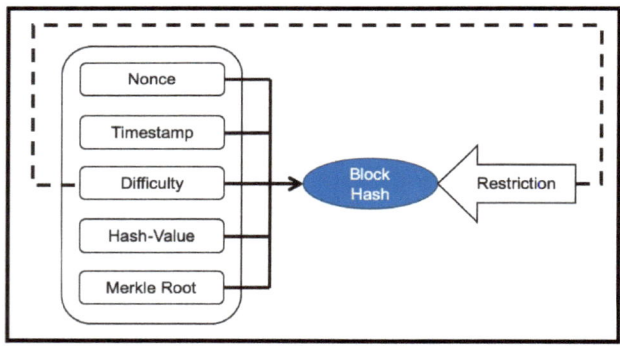

Fig. 4.5: Schematic representation of a hash puzzle. Source: Drescher (2017): p. 156.

In addition, a **nonce** (freely changeable data) must be included, which is required to solve the hash puzzle. When solving the hash puzzle, please note that the hash value of the block header must meet the specified restrictions or **difficulty level**. Since the degree of difficulty is a part of the block header, it also flows into the hash value of the block. In this way it is ensured that nobody can avoid the computational effort of a hash puzzle by deliberately reducing the difficulty level. The central concern of the hash puzzle is to impose a certain restriction on the hash value. Hash puzzles can only be solved by trial and error. This requires **guessing a nonce**, **calculating the hash value** of the combined data with the required hash function, and evaluating the resulting hash value based on the constraints. The nonce, which in combination with the existing data results in a hash value that complies with the constraints, is called the solution. Only blocks whose headers contain a correct solution for their respective hash puzzle are processed further. Any block whose header does not pass a proof of work check will be discarded immediately. If a block is discarded, continue with another Nonce until the puzzle is solved. To prove that the hash puzzle has been solved, the corresponding nonce must always be delivered. The type of restriction used for hash puzzles is standardized so that computers can challenge each other with hash puzzles. In this context, restrictions are often referred to as **difficulty levels**. Difficulty is expressed as a natural number and indicates the minimum number of leading zeros the hash value must have. For example, if the difficulty is 1, it must have at least one leading zero. If the difficulty is e.g. 20, the hash value must have at least 20 leading zeros. The following applies: The higher the difficulty level, the more leading zeros are required and the more complicated the hash puzzle is. As complexity increases, so does the amount of computing required or the time required to solve the puzzle (cf. Drescher 2017, p. 107ff.). If the puzzle is solved too quickly, e.g. because the Miner's hardware or software has been optimized, the difficulty level is automatically equalized. With the Bitcoin, for example, this ensures that the interval of ten minutes between the creation of new blocks is not undercut (Rosenberger 2018, p. 68). In principle, the solution to the cryptographic task can be illustrated using the flow diagram shown in Fig. 4.6.

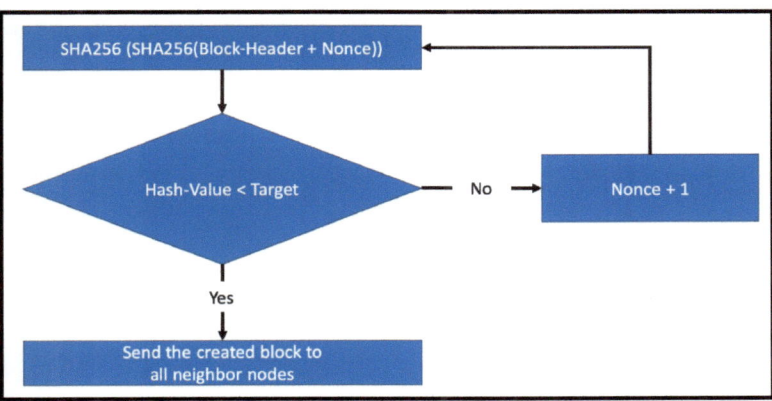

Fig. 4.6: Flowchart for solving a hash puzzle. Own representation based on Meinel et al. (2017): p 41.

In this example, the hash value is calculated by the double SHA-256 hash function from the block header and a nonce. The nonce is adjusted until the hash value is less than the target. The degree of difficulty increases with the number of leading zeros in the target. To illustrate how this works with numbers, the following example describes an example in which the target requires at least three leading zeros. Table 4.1 below lists the nonce, the text for the hash function, and the resulting shortened hash value. Since the target contains at least three leading zeros, the puzzle is solved with Nonce "89". After the calculation has started with a nonce of zero, the nonce is increased by the value "1" with each lap in which the target was not reached, as shown in Fig. 4.6.

Table 4.1: Example for solving a hash puzzle. Own representation.

Nonce	Text for the Hash function	Output
0	Energiewende! 0	NJNKCS8Z3WFNC3C382C3544NKJNR49A
1	Energiewende! 1	SDBD28E21ENDJKNJKWNQWR98390IDG
2	Energiewende! 2	34F748HFNU43FF4309I3F32FD33DWDQQ
3	Energiewende! 3	0NJSDND72ZBHB278G21DN5453JGJHJFF
…	…	…
89	Energiewende! 89	000HU7Z237Z2NDWQ87124TSF4545FEWE

In this case, 90 runs or trials are required to find the solution (000HU7Z237Z2NDWQ87124TSF4545FEWE). If instead a hash value with only one leading zero were required, the task would already be solved after four steps, since the entry "Energy Transition! 3" yields a hash value with a leading zero (0NJSDND72ZBHB278G21DN5453JGJHJFF). Depending on the validation result, the block is accepted or rejected. If the validation is successful, the created block is distributed to all other participants.

Creation and Linking of New Blocks

The **creation** and **concatenation** of new blocks forms another core element of the blockchain. The first block of a chain (Genesis block) has no reference to previously executed transactions. The next block with new transactions is appended to a chain with past blocks consisting of individual transactions. In principle, each block refers to its predecessor by containing the hash value of the predecessor block. The following example illustrates how the entry of new transactions or the concatenation of blocks takes place. Fig. 4.7 shows the initial situation of a blockchain data structure, which so far only consists of one block and two further transactions.

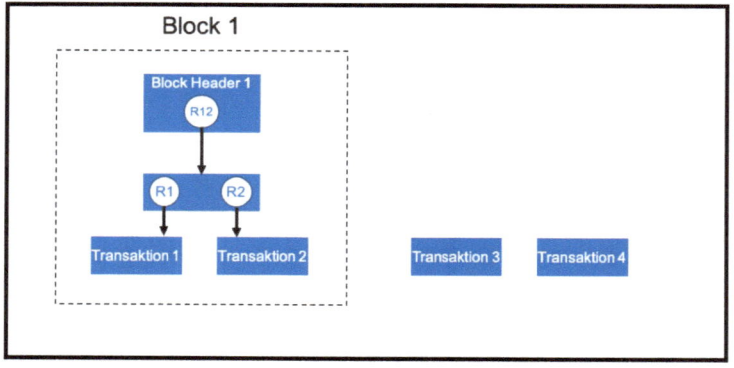

Fig. 4.7: Initial Situation. Source: Own Representation Drescher (2017): p. 142.

It is assumed that transactions 3 and 4 have not yet been added to the blockchain data structure. To achieve this, a new hash tree must first be created that contains all new transaction data to be entered (Fig. 4.8).

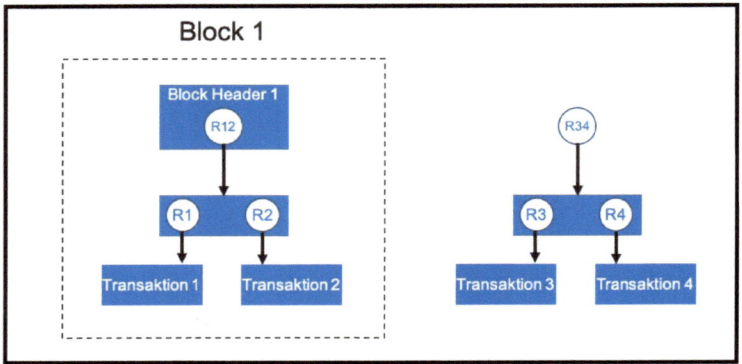

Fig. 4.8: Creation of a New Hash Tree. Source: Own Representation Drescher (2017): p. 143.

After the new hash tree has been created, a new block header (Block Header 2) must be created (Fig. 4.9). This contains the hash reference B1, which refers to the header of the previous block (Block Header 1). In addition, block header 2 contains the root of the hash tree of block 2, which contains the new transaction data (R34).

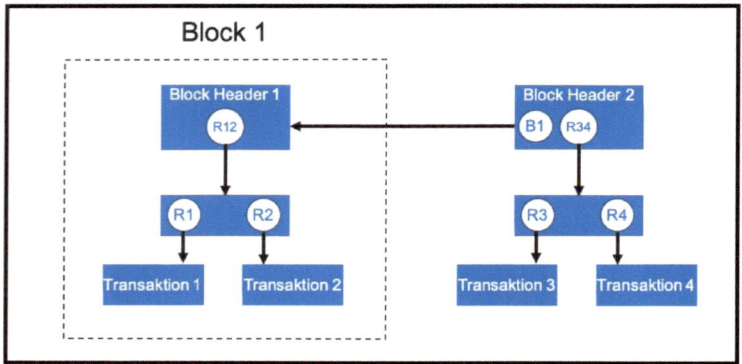

Fig. 4.9: Creating a New Block Header. Source: Own Representation Drescher (2017): P. 143.

For a new block to be connected to the existing chain, consensus building as described above is essential. After successful validation, the new block 2 forms the reference for subsequent blocks. To add another block in this example, a new hash reference (B2) must be created, which in turn refers to the last validated block header 2 (Fig. 4.10).

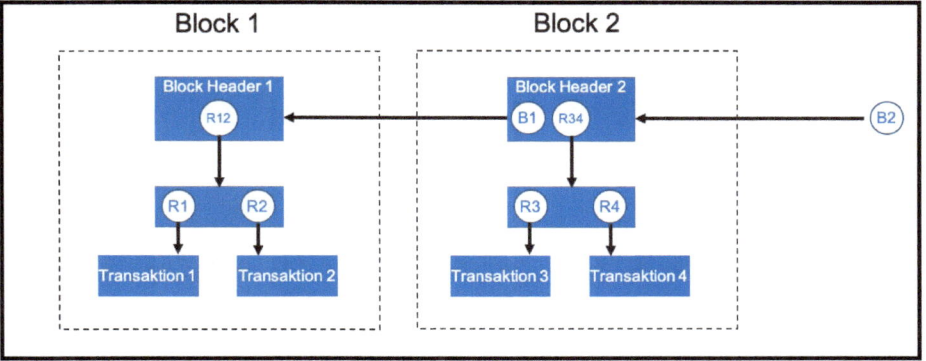

Fig. 4.10: Creating a New Hash Reference. Source: Own Representation based on Drescher (2017): p. 143.

Due to the unique concatenation, the blockchain can be regarded as tamper-proof, since even a small change in a single statement would change the hash value of the block and the hash tree would therefore no longer be consistent. (Schütte et al. 2017, S. 10).

4.3 Classification of Blockchains

Basically it must be made clear that there is no "one" blockchain. The term blockchain refers only to the system of data storage using encrypted chaining. The design possibilities of a blockchain are manifold and concern different details. A distinction can be made between the access to the data and the use of the network, as well as the possibilities to participate in the validation of blocks in the consensus mechanism. While with a "public blockchain" everyone has access (public), with a "private blockchain" access is restricted (private). Regarding participation in the consensus mechanism, a "permissionless blockchain" allows anyone to participate and write new blocks into the blockchain, while a "permissioned blockchain" allows only authorized nodes to check transactions and write new blocks. Hybrid solutions include shared permissioned blockchains, which are a compromise between public and private blockchains (consortial).

Public Blockchain

In the case of public blockchains, access to the blockchain is free for each user. Since no proof of identity is required, public blockchains belong to the "permissionless blockchains". This allows each user to participate in the network and validate transactions. Public blockchains so far are mainly based on the POW consensus mechanism for creating new data blocks. In principle, the POS consensus mechanism can also be used for public blockchains. However, this requires a certain minimum deposit in order to validate transactions, which results in a form of access restriction to the validation of transactions (Bogensperger et al. 2018a, p. 13).

Private Blockchain

A private block chain is in most cases a permissioned block chain, which means that only selected participants may validate the transactions (Schütte et al. 2017, p. 9). The access rights or permissions are managed by one or more central instances. Read and write permissions can be assigned at will. The difference between a private and a public blockchain is also the degree of decentralization and the guarantee of anonymity (Bogensperger et al. 2018a, p. 14).

Consortia Blockchain

Consortial blockchains are a compromise between public and private blockchains. While in a private blockchain the validating nodes are only operated by one legal entity, in a consortium blockchain this task is distributed among a consortium of several organizations in the network. Typically, the consensus for this blockchain variant is achieved by a majority decision of the authorized participants (Bogensperger et al. 2018a, p. 14). This limits the scope of consortium blockchains, since both the participating nodes and the approved applications require the approval of the entire consortium. On the other hand, this tested approval is highly attractive for companies (Bundesverband der Energie- und Wasserwirtschaft 2017, p. 20). This is why they are frequently used by financial service providers and in the energy industry. Examples are *"R3"*, *"The Energy Web"* and *"The Blockchain Insurance Initiative"*. In principle, consortia blockchains can also be used for certain tasks in public administration, for which, for example, the outstanding reliability and transparency of public blockchains should be dispensed with for data protection reasons (Blocher 2018, p. 11).

4.4 Strengths of Blockchain Technology

In principle, blockchain technology offers versatile strengths. These include above all the following strengths, which will be explained in more detail below:

- Disintermediation
- Data Integrity
- Safety
- Trust
- Transparency
- Automation
- Cost Reduction

- Anonymity

Disintermediation

Up to now, the continuous digitalization of processes has required that the players know each other and trust each other, and that liability is assumed by third parties. For the first time, a blockchain enables the direct and secure processing of digital transactions between unknown players without the need for a mediation function. With the blockchain, trust is generated by consensus mechanisms. For this purpose, all full nodes involved in the network check and confirm the correctness of the transactions. Since the trust authority is formed by the decentralised community, intermediaries become superfluous (disintermediation) (Bundesverband der Energie- und Wasserwirtschaft 2017, p. 23). Drescher points out that the role of the mediating authority or the mediator in a block chain is not completely eliminated, but is substituted (Drescher 2017, p. 251f.). In principle, it should be borne in mind that the use of blockchain technology and the associated substitution of intermediaries is only expedient if there is no trust in third parties (cf. Eckstein et al. 2018, p. 70).

Data Integrity

Due to the cryptographic linking of the blocks, a very high degree of integrity of the data in the blockchain can be ensured, since any change to an identity characteristic of a block results in a change to the characteristics associated with it. In order to make a proper change within the blockchain, all hash references would need to be updated, while the time and cost required for such an operation also increases as the length or size of a blockchain increases. As a result, the transaction history in the blockchain is virtually unchangeable. This property is mandatory for intermediation-free transaction processing and for the decentralized distribution of verification and validation processes (Fischer et al. 2019, p. 452). Since the transaction history in the blockchain cannot be changed without further effort, this aspect also contributes to the security of a blockchain and thus forms another advantage.

Safety

Data integrity allows stored transaction data in the blockchain to be checked for authenticity, origin and integrity. This feature makes the blockchain a secure storage medium for sensitive data or the status of data and documents. The decentralization of data processing and the database itself is also central to the use of a blockchain. The database is stored redundantly by all or some of the participants (Fischer et al. 2019, p. 454). In view of the fact that *"both the blockchain itself and individual mechanisms such as the verification of digital signatures (...) are often reproduced by all network participants, there is no single point of failure within the network"* (Schlatt et al. 2016, p. 35). Thus, the resilience of the network can be significantly increased and the data and information can be protected against server failures and attacks. The following applies: the longer a block is contained in the blockchain, the less likely it is that an attacker will manage to generate the same hash value on the basis of manipulated transaction content and thus make changes in the network (Bundesverband der Energie- und Wasserwirtschaft 2017 p. 24f.).

Trust

Due to the security of the system, trust in intermediaries is not necessary. Any interaction is therefore free of uncertainty or fear of fraud. What used to be the task of intermediaries such as governments, banks or corporations is now being taken over by blockchain technology (Cryptolist, no year). According to Palka and Wittpahl (2018), a distributed P2P network creates additional trust, especially in conjunction with consensus mechanisms.

Transparency

Since each new transaction is recorded and distributed across many computers, each change can be tracked in the blockchain. This transparency enables the documentation and insight of processes. This possibility can be a decisive advantage especially for different actors who need a common database (Bundesverband der Energie- und Wasserwirtschaft 2017, p. 24). The high degree of transparency ultimately also enables intermediation-free verification and processing of transactions and thus a system in which third-party network participants can mutually control each other (Fischer et al. 2019, p. 453).

Automation

According to Drescher, manual tasks can be replaced by automated interactions with increasing use of the blockchain, which is why the potential of the blockchain to promote automation can be counted as another advantage (Drescher 2017, p. 252). In addition to transactions, the flow and execution of a computer program with process logic within an organization can also be recorded in a blockchain. Further design options are possible through the integration of Smart-Contracts. By their automated execution a higher contract security can be ensured, since subsequent action deviations are practically not possible. This enables completely new forms of organization that allow automated transactions to run. Automation should also reduce transaction costs, which is the next advantage (Bundesverband der Energie- und Wasserwirtschaft 2017, p. 24).

Cost Reduction

According to Drescher, the economic consequences of disintermediation and automation are often reflected in falling costs. Cost reduction in the initiation and execution of transactions could prove to be the most sustainable contribution of the blockchain from an economic point of view (Drescher 2017, p. 253). Since a blockchain can assume the role of an intermediary as a protocol, only very low costs are incurred compared to a central unit. In addition to increased profitability in existing markets, this also allows new markets to be defined and opened up that were previously unavailable for cost reasons (Fischer et al. 2019, p. 454).

Anonymity

Anonymity is a central factor especially in public blockchains. Strictly speaking, it is pseudoanonymity (Bundesverband der Energie- und Wasserwirtschaft 2017, p. 24). Although the participants themselves are represented by cryptographically encrypted number series, any access and activities of the participants can be transparently traced. Depending on the type of blockchain, it is very difficult or even impossible to trace the real identity of a participant (Cryptolist, no year).

4.5 Weaknesses of Blockchain Technology

However, the structural features of the technology also entail various problems or weaknesses. These problems can vary with different blockchain implementations. The following weaknesses are seen in detail:

- Scalability
- Energy Consumption
- Irreversibility
- Data Protection
- Legal Framework

Scalability

Perhaps the most significant weakness is the scalability of the blockchain solution. In general, the scalability of a system is understood as *"(...) its ability to improve with the growth of resources"* (Malanov, 2017). Public blockchains in particular are not yet mature enough to handle a large number of transactions (Fischer et al. 2019, p. 455). The two leading blockchain applications, BTC and ETH, operate with limited block sizes. The rule here is: the more transactions, the more data the individual blocks must carry. This in turn can lead to overloading (Alexandre, 2018). Since the speed of transactions cannot be influenced by adding new resources, the classic blockchain is not scalable. According to Sinegal, the scalability *"(...) remains a big problem of the blockchain. Typically, all nodes process all transactions in parallel. This leads to enormous inefficiencies"* (Sinegal, 2018).

Energy Consumption

This weakness relates in particular to the elaborate calculations (mining) which are carried out within the framework of the POW consensus mechanism and thus cause enormous energy wear which is not compatible with environmental protection (Egloff and Turnes 2019, p. 114). If ten percent of the world's population were to use blockchain technology in this way, around 23 percent of the world's electricity would have to be produced in cold ovens and cans (Kalthofen und Can, 2018). However, there are alternative consensus mechanisms such as the POS, which works without high computing costs and is therefore energy-neutral. For this reason, according to Florian Glatz, President of the Blockchain Bundesverband, most new blockchain projects no longer use a POW process to secure their infrastructure. Glatz predicts that the POW consensus mechanism will ultimately be limited to Bitcoin, while all other networks will switch to the energy-neutral POS consensus mechanism (Glatz, 2018).

Irreversibility

Irreversibility follows from the POW consensus procedure, which must be carried out every time the data of a block is changed. Although this property is an essential part of data integrity, incorrectly entered and sent transactions cannot be reversed (Palka and Wittpahl 2018, p. 11f.). This is particularly critical for Smart-Contracts, as errors in the program code cannot be changed afterwards. In addition, it is conceivable that "backdoors" are programmed into the program code, whereby the reliability and immutability of a Smart-Contract based on a blockchain depends on the respective programmer (Schiller, 2018b).

Data Protection

Transaction processing without intermediation lags behind the regulated payment infrastructure of the traditional financial system with regard to the regulatory framework. On the one hand, the irreversibility or immutability of data described above is in direct conflict with, for example, the rights of data subjects to have data corrected or deleted and the right to be forgotten under Art. 17 of the EU Services Regulation. On the other hand, the particularly high degree of transparency in public blockchains has a negative effect on data protection, as the transaction history can be viewed by any participant. This circumstance can, however, be circumvented with private blockchains by only giving access to authorized participants. Private blockchains can also provide a remedy with regard to the immutability of data by enabling a change to be decided retrospectively by a consensus of all participants or, in particular, by a consensus of the validators (Fischer et al. 2019, p. 452f.). According to Drescher (2017), the aforementioned conflicts could trigger users to "(...) become aware of the importance and economic significance of their personal data and to claim ownership of it more strongly than before".

Legal Framework

In addition to these more technical risks, BaFin also sees supervisory and legal risks. Since blockchain implementations basically function without the borders of nation states, two transaction participants may be located in different jurisdictions. This is particularly clear in the case of implementations of public blockchains, where conflicting legal regulations can lead to confusion as to the applicable or valid regulations. Since the legal significance has not yet been clarified either in a blockchain transaction or in a smart contract, answering these fundamental questions still involves a certain risk factor at the present time. Since participation in implementations based on private blockchains is linked to the acceptance of certain legal rules, it would potentially be easier to eliminate these uncertainties. It should be borne in mind that even when blockchain technologies are used, the existing regulatory framework within BaFin's area of responsibility applies if the parties involved are subject to BaFin's supervision. Due to the fact that only the supervisory facts form the starting point for BaFin's supervisory work and that therefore there is no restriction of the blockchain technology. Therefore, it is not the technology but the use case that is decisive for regulatory issues. BaFin further states that difficulties in applying supervisory law only arise if, in the absence of a central authority, its enforceability would be made more difficult or impossible due to a lack of addressees (BaFin, 2017).

5 Blockchain in the Energy Industry

Many experts believe that blockchain technology can bring about far-reaching changes in the energy sector (Bundesverband der Energie- und Wasserwirtschaft 2017, p. 5). By far the most discussed use case is due to P2P trading, which is intended to make electricity trading between private individuals without a participating energy company, and thus also for prosumers despite low output, economically possible. The most important arguments for the use of blockchain technology in the energy market include simplified and automated processes, greater transparency and the reduction of transaction costs through disintermediation. However, there are also some arguments against the use of blockchain technology in the energy market. These include in particular the low transaction speed, illegal activities as well as the consumption of energy and resources. In addition, there are fundamental legal issues such as data protection. According to many estimates by experts from the energy sector, blockchain technology has the potential to significantly influence the energy industry in the coming years and thus unfold a new dynamic for the energy transition.

5.1 Assessments by Experts from the Energy Industry

An initial impulse for what the blockchain can achieve in the energy industry was provided by a brief study commissioned by the NRW Consumer Centre in 2016 by the management consultancy PricewaterhouseCoopers (PwC). According to this study, the developments to date *"(...) show that Blockchain can strengthen the role of individual consumers and producers in the market in the future. Prosumers are given the opportunity to trade the energy they generate directly with a high degree of independence via Blockchain technology. Blockchain technology therefore promotes the development towards further decentralisation of energy systems"* (Hasse et al. 2016, p. 39). For Marten Bunnemann, CEO of Avacon AG, the importance of data and its digital use increases considerably with the increasing decentralisation of the energy world. Therefore, according to Bunnemann, the implementation of municipal climate protection goals should also take place via digital solutions (Bunnemann, 2018). According to Thon et al. (2018), authors of the study „Blockchain – eine Technologie mit disruptiven Charakter", *"(...) the use of blockchain technology for the digitisation and decentralisation of the energy system transformation is currently without alternative"*. Maier puts forward the thesis that blockchain technology is the most frequently discussed aspect of digitalization in the energy industry. According to him, smart contracts in particular could be used to digitally and automatically process transactions in electricity trading between decentralised electricity producers and consumers and thus replace third parties such as traditional energy suppliers, banks and exchanges. Since blockchain technology is still at an early stage of development and many questions still need to be clarified, Maier believes that such an application can currently only be recognised as a vision (Maier 2018, p. 8). In a survey conducted by EUWID, Dr. Sönke Gödeke, attorney at law at Pinsent Masons with a special focus on the energy sector, commented on the prospects of the blockchain in the energy industry. In his opinion, blockchain technology has the potential *"(...) to significantly influence the energy industry in the coming years"* (Preiß, 2018a). Doleski (2017) is also certain that blockchain technology will play a major role in the upheaval of energy supply. According to Benjamin Talin, entrepreneur and consultant in the field of digitalization, the transparency and traceability of blockchain technology can achieve great progress, especially in the complicated energy market. Talin believes that the positive properties of blockchain technology could have a major impact on a successful energy transition (Talin, 2019). According to Andreas Kühl, author of one of the best-known and most influential energy blogs, there is no way around the prosumer. For Kühl, the trend towards local power generation, storage and use, and thus towards prosumers, is clearly discernible. In this context, Kühl mentions the blockchain technology that could be used in the future to enable electricity trading in the neighbourhood (Kühl, 2017). According to Prof. Dr. Dr. Walter Blocher, Professor at the Institute for Commercial Law at the University of Kassel, the energy industry is *"(...) primarily about a foundation for decentralised energy supply, which is hardly conceivable without blockchain-based billing methods"* (cf. Blocher 2018, p. 6). For Angelika Ehrlich, editor of Energie & Management GmbH, the blockchain plays an important role regarding business models in the digitalization of the energy industry and the energy transition (Ehrlich, 2017). As the world's first energy provider of a decentralized home storage system for blockchain-based energy services, "Sonnen GmbH" sees the blockchain as the next evolutionary step in decentralized energy supply. Sonnen GmbH assumes that the blockchain technology is the only solution for the networking and economical power exchange of many individual systems associated with decentralised energy supply (Sonnen GmbH, 2017). For Michael Lucke, managing director of the Allgäuer Überlandwerk, *"(...) decentralisation, decarbonisation and digitisation (...)"* are the central challenges of today's energy market. According to Lucke, blockchain technology can help to overcome the challenges mentioned. Lucke thus also suspects an alternative marketing opportunity for owners of PV systems when the EEG remuneration for the first systems expires in the next few years: *"They could offer their green electricity on this platform, while consumer-conscious consumers can purchase their electricity directly from their neighbourhood or region"* (Kloth, 2017a). Siegel and Andersen, partners and heads of a renowned auditing firm, assume that the energy sector will be confronted with disruptive changes particularly quickly by blockchain technology: *"Consumers will sell the electricity produced, for example, with their own solar system independently of energy suppliers"* (Siegel and Andersen, no year) According to a study by BDEW from 2017 with the topic

"Blockchain in der Energiewirtschaft", the energy sector will receive a highly interesting technology with the blockchain, with which a secure, decentralised and flexible exchange of information can be made possible. The blockchain technology should have the potential to *"(...) serve as a transformer for a digitally networked ecosystem of millions of devices"*. In addition, the development of ever new applications based on blockchain technology and numerous projects by energy suppliers underline the high dynamics and the associated expectations. In addition, the study concludes that energy supply companies can secure new roles and segments in this early phase by actively participating in the development of the blockchain. However, in addition to these opportunities, technical limitations should also be analysed or whether existing IT solutions offer similar advantages. In addition, legal issues relating to blockchain technology must also be considered (Bundesverband der Energie- und Wasserwirtschaft 2017, p. 5). As a result of a further study by dena and the European School of Management and Technology (ESMT) in 2018 on the subject of *"Vulnerabilities in smart meter infrastructure - can blockchain provide a solution?"*, Burger et al. (2018) conclude that the use of smart meters together with decentralised transaction technologies such as blockchain can be used for direct trade between producers and consumers in P2P networks, which in the longer term could even call into question the intermediary and coordination role of energy supply companies in certain market segments. According to Kloth, this is also one reason why *"(...) more and more energy companies are experimenting with the blockchain: They want to develop their own offers in good time so as not to be overwhelmed by the technology"*. Kloth sees the blockchain technology as a hope for the decentralised energy transition, with which direct P2P trading of electricity can be made possible, e.g. between the owner of a solar plant and his neighbour (Kloth, 2017b). Sebnem Rusitschka, managing director of *"freeel.io"* and head of the energy working group in the Blockchain Bundesverband, demands *"(...) a Blockchain-capable infrastructure, for example for the Smart Meter Gateways"* and that the requirements of the regulatory authorities can be adapted to new technologies. According to Rusitschka, safety solutions for the blockchain in smart meters are available (Kloth, 2018a). According to Carol Inoue Dick, Business Developer at Vattenfall Trading, *"besides more efficient and cheaper transactions in electricity and gas trading on the basis of blockchain technology, decentralized generation plants could be integrated into the grid much more easily in the future"*. From Dick's point of view, this would be an important contribution to the expansion of renewables. According to Dick, the decentralized and tamper-proof storage of transaction data would also make it possible to provide proof of authenticity for electricity from renewable' energy sources (Hannen, 2018). According to Franz Nees, Dean of the Faculty of Computer Science and Business Informatics at Karlsruhe University of Applied Sciences, the blockchain could become an enabler for a decentralised energy supply system. Nees believes that blockchain

technology can potentially greatly simplify the traditional system from power generators, TSOs, VSOs to consumers by connecting consumers and producers directly. According to Nees, the new system could create a possibility for prosumers to transmit transactions between business partners directly via the P2P network using blockchain technology. According to Nees, in addition to P2P trading, the blockchain can also be used to control networks through smart contracts and to map ownership relationships with regard to the electricity itself, but also with regard to the CO_2 certificates and green certificates required for this (Nees, 2018). According to Dr. Ansgar Steinkamp, Senior Expert at Open Grid Europe, blockchain technology could enable energy producers and consumers to *"(...) directly exchange energy ownership without the intermediaries and intermediaries that are common today. The participants could then interact directly with each other via the blockchain and trade in electricity, gas and certificates for renewable energy"*. According to Bastian Wilkat, Digital Strategist at BTC Business Technology Consulting AG, the blockchain technology would make it easy to connect decentralised generation plants to the power grid and promote the production of renewable energy to support energy system transformation. *"The blockchain could trigger major upheavals in the energy industry,"* says Wilkat. Proof of authenticity for electricity from renewables or CO_2 certificates can also be realized with the blockchain, which is attributable to the ability to store transaction data in a decentralized and tamper-proof manner (Schreier, 2017). For Robert Doelling, an expert in renewable energy and energy technology, blockchain technology is particularly predestined for commercial transactions in which only small quantities of electricity are supplied from the producer to the consumer. *"One of the probably most important financial arguments in favor of the blockchain in the power grid is that it could reduce the high and heavily regulated transaction costs in the German power industry to a minimum,"* says Doelling. Doelling (2016) concludes that blockchain technology has the potential to make a decisive impact on energy system transformation. According to Rüdiger Winkler, coordinator of the Blockchain-Initiative Energie and managing director of the EDNA-Bundesverband Energiemarkt & Kommunikation, *"hardly any other term in the energy industry has received as much attention as the word "blockchain" in recent years, even though the term was not directly taken into account in GDEW in 2016"*. Winkler is certain that *"(...) this technology will probably also have serious consequences for the energy industry"* (Winkler, 2019). The Federal Government has also recognised this and, with the coalition agreement, has committed itself to developing a comprehensive blockchain strategy to provide constructive support for the development of the technology (Federal Government, 2018a). In the view of the Federal Government, the blockchain is *"(...) a potential new basic technology for digitisation, the technological properties of which can open up a broad, cross-sector field of possible applications"*. In the view of the Federal Government, strategic monitoring of this development is necessary at this early stage of the technology in order to strengthen the competitiveness and innovation capability of the German economy and to secure technological sovereignty. With regard to potential energy industry applications, it should be critically examined whether they make sense with regard to the national and international climate targets as well as in terms of energy transition and security of supply. The publication of the blockchain strategy is planned for summer 2019 (German Bundestag 2019, p. 1). During the meeting of the working group „Zukunftsenergien" in the Forum for Future Energies on 30 January 2019 on the topic „Blockchain – Die Lösung für eine dezentrale Energiewende?" it was discussed in which areas of the energy industry this technology can be used and which regulatory framework is required for this. For Philipp Richard, Team Leader for Energy Systems and Digitisation at dena, digitisation is an indispensable building block for the transformation of the energy system. According to Richard, the classic value chain is developing into a value creation network with an information exchange of millions of different assets. Overall, the system complexity will increase as a result of digitalization, but the opportunities will clearly outweigh the risks, according to Richard. According to Richard Plum, Product Manager Consulting at ProCom GmbH and Chairman of the Blockchain-Initiative Energie in the EDNA-Bundesverband Energiemarkt & Kommunikation e.V., the Blockchain offers some advantages in direct comparison to classical networks. For example, a high degree of automation can be achieved using smart contracts, which leads to a reduction in costs. Winkler also emphasized that the blockchain represents a solution for a decentralized energy world of the future. In the current system, the blockchain could already provide a solution for the marketing of such plants that fall outside the scope of EEG funding. Due to the current regulatory framework, however, it does not yet make sense to use the blockchain across the board, Winkler said. According to Dr. Torsten Kraul, Associated Partner and Attorney at Law at Noerr LLP, there is currently no suitable regulatory framework within which blockchain technology can be operated in a cost-covering or profitable manner. For Kraul, politics has a duty to develop this regulatory framework. Timon Gremmels, Member of the Bundestag (SPD), said that the upcoming blockchain strategy on the part of the federal government would create a "tool" for the energy transition. All in all, Gremmels sees a high cost reduction potential in digitisation, which can be attributed to the efficiency gains. Prof. Dr. Martin Neumann, MdB (FDP), also underlined the potential of the blockchain, especially in view of the increasing need for coordination within a decentralized energy world. According to Neumann, the blockchain strategy is the first right step to give the industry planning security and trust. However, a "blockchain law" must follow in order to guarantee this, said Neumann. Dr. Ingrid Nestle, MdB (Bündnis 90/Die Grünen), supports the development of digital solutions, but currently sees no particularly "revolutionary" use cases specifically for blockchain technology. However, it is important that the federal government creates a framework that allows, among other things, the blockchain to be used sensibly (Zunkunftsenergien 2019, p. 1f.). In an earlier article in the Handelsblatt, Richard described that the blockchain could help to bring together the large number of producers with a large number of customers. *"Blockchain technology is not just a hype, it can really help to boost the energy revolution,"* says Richard (Witsch, 2018). For Andreas Kuhlmann, Chairman of dena's Executive Board, the second phase of the energy revolution will also be *"(...) about "intelligently connecting the various components and players in the energy system"*. According to Kuhlmann, blockchain technology offers a promising approach: *"It is a good sign that some pioneers in the energy industry are dealing constructively with this trend. Whether it will lead to successful business models cannot yet be said today. But energy system transformation needs innovative players who want to shape the future courageously"*, Kuhlmann continues (dena, 2016). In 2016, Kuhlmann et al. (2016) together with ESMT conducted a survey of 70 experts from the energy industry as part of the study on „Blockchain in der Energiewende", the results of which are presented below.

Survey on the Study „Blockchain in der Energiewende"

In July and August 2016, dena and ESMT interviewed decision-makers from the German energy industry about their assessments, existing and planned activities and their visions on blockchain. The survey was answered by a total of 70 executives along the entire value chain of the energy industry. The survey starts with the general question whether the participants have already heard of blockchain applications in the energy sector. Over two thirds (69%) of respondents said they had already heard of blockchain applications in the energy sector. Approximately one third (31%) have not yet heard of blockchain applications in the energy sector (Fig. 5.1).

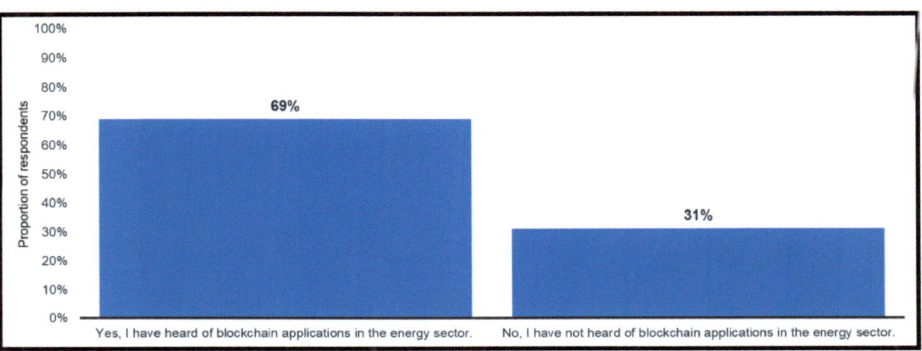

Fig. 5.1: Result of the Survey on the Awareness Level of the Blockchain in the Energy Sector. Source: Cf. Kuhlmann et al. (2016): p. 17.

In the second question, the decision-makers were asked about the degree of implementation of the blockchain in their companies or organisations (Fig. 5.2).

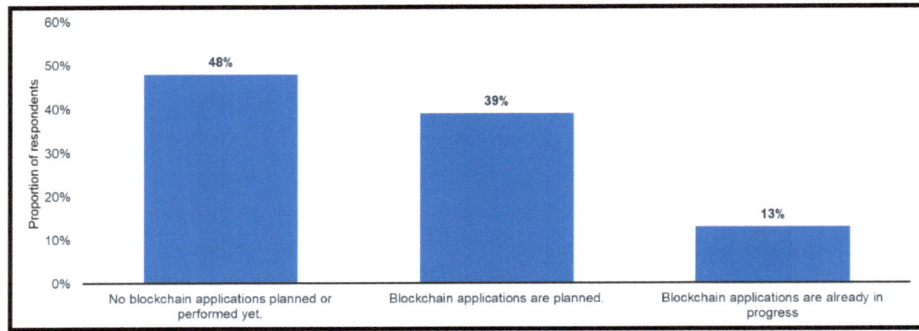

Fig. 5.2: Result of the Survey on the Degree of Implementation of the Blockchain in the Energy Sector. Source: Cf. Kuhlmann et al. (2016): p. 18.

Almost half (48%) of all respondents stated that blockchain activities had not yet been carried out or planned in their companies or organisations. However, 39% of those surveyed are planning blockchain applications in the form of pilot projects, studies, analyses and research projects. Only about 13% of the respondents already carry out activities regarding blockchain applications, such as active scouting of start-ups, business development or dealing with the blockchain element "proof-of-concept". Beyond the current applications, the participants were asked about their assessments of the potential of the blockchain in the energy sector (Fig. 5.3).

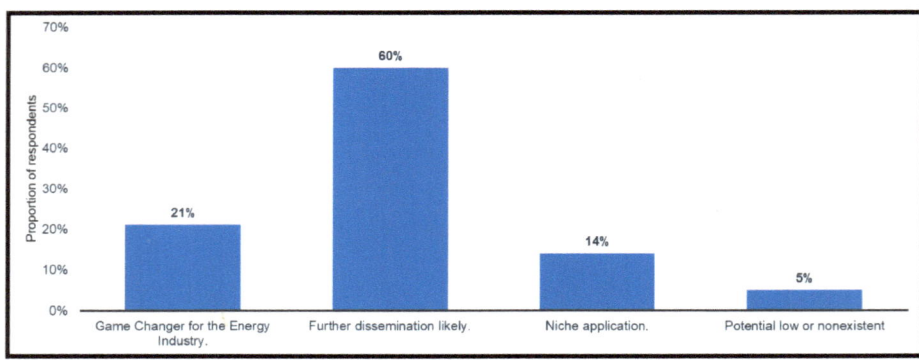

Fig. 5.3: Result of the Survey on the Potential of the Blockchain in the Energy Sector. Source: Cf. Kuhlmann et al. (2016): p. 19.

More than half (60%) of the decision-makers believe that the blockchains are likely to spread further in the energy sector. A further 21% see the blockchains as game changers for the energy sector. On the other hand, 14% of the survey participants expect the technology to remain limited to niche applications. For 5 % of the respondents, the blockchains have no potential in the energy sector, or only one that can hardly be mapped. Furthermore, the decision-makers participating in the survey provided information about possible examples of blockchains in the energy sector. Overall, the respondents cited more than 110 possible use cases and estimated the respective potentials. The answers were divided into processes and platforms. According to the interviewees, blockchain technology could prove itself as a "game changer" in the field of security. However, this application case is considered potentially uninteresting due to the lowest answers. In the area of platforms, P2P trading, such as direct electricity trading between owners of PV systems and consumers, was classified as "likely to spread further" and, on the basis of the most frequent responses, as the application with the greatest potential. The results of the survey suggest that German energy executives see a wide range of possible blockchain applications in the energy system, both in terms of processes and platforms. According to the responses of the decision makers, blockchains had the potential to positively influence the energy sector by reducing costs and creating new business models or marketplaces. Particularly noteworthy is P2P trading, e.g. between prosumers, which was rated as the application with the greatest potential in the energy sector. The answers suggest that blockchains can become an important building block for the transition to energy system transformation 2.0 (Kuhlmann et al. 2016, p. 23).

Survey on the study „Blockchain in der integrierten Energiewende"

The survey was conducted as part of dena's study on „*Blockchain in der integrierten Energiewende*", which was published on 26 February 2019. 300 managers and experts from the energy industry in Germany, Austria and Switzerland took part in the survey. First, the participants were asked about the stage of their companies' involvement with Blockchain (Fig. 5.4).

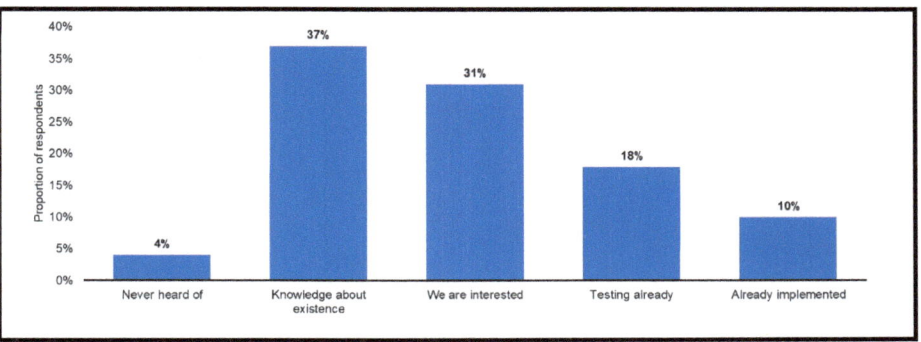

Fig. 5.4: Stage of preoccupation with blockchain. Source: German Energy Agency (2019): p. 2.

More than one third (37%) said they knew about the existence of the blockchain. This makes it clear that the topic of blockchains is widely known in companies in the energy industry.

Another 31% of the companies surveyed were interested in blockchain technology. More than a quarter of the companies surveyed stated that they had experimented with the blockchain in various energy use cases (18%) or even had already implemented it (10%). Only 4% said they had never heard of blockchain technology. A further survey focused on the blockchain use cases of the companies surveyed (Fig. 5.5).

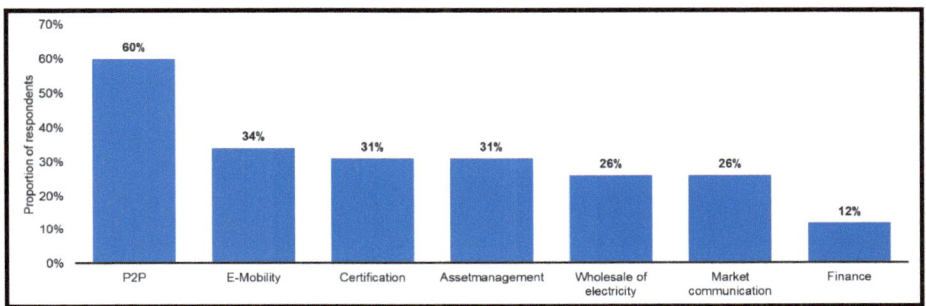

Fig. 5.5: Blockchain use cases of the surveyed companies. Source: Cf. German Energy Agency (2019): p. 6.

P2P trading, in which market players want to act directly with each other without central instances such as stock exchanges, brokers or energy providers, was given as by far the most frequently stated use case with 60%, which is tested by the companies surveyed on the basis of blockchain technology. E-Mobility was rated at 34% as the second most common use case on which the companies surveyed experimented with the blockchain. In each case, 31% of the use cases are limited to certification and asset management. Applications in the area of electricity wholesale and market communication were each given with 26 %. At 12%, use cases in the finance sector are the least tested within the framework of companies on the basis of blockchains. In addition, the companies were asked to what extent there is a connection to knowledge management and competitive pressure in research on blockchain technology. Since these criteria play a subordinate role for the present work, reference is made to the German Energy Agency (2019) for the presentation of the results and a more detailed explanation. The results of the survey show that more than a quarter (28%) of the companies surveyed are experimenting with the blockchain in different energy use cases or have even already implemented it. P2P trading is by far the most common application area, which is tested or implemented by companies on the basis of blockchain technology. According to Kuhlmann, *"(...) several companies are using, developing or even offering the new technology on the market. In addition to digital technology competence and a feel for business ideas, the corporate culture seems to have a high influence here"*. Kuhlmann concludes that the blockchain can become a real driver of energy system transformation if the above factors come together (dena, 2019a).

5.2 Opportunities and Challenges

As part of a meta-analysis of the AEE on the role of digitisation in the energy system transformation process, the statements of scientists on the opportunities and challenges of blockchain technology were aggregated from nine different studies, which are explained in more detail below.

Opportunities

In principle, the opportunities that could result from the use of blockchains in the energy industry can be divided into economic, technical and social aspects. From an economic point of view, the potential **reduction of transaction costs** is a particular argument in favour of the technology, since, for example, detours via intermediaries could be avoided by means of program codes. Since many small-scale transactions via today's central trading platforms are too complex and cost-intensive in relation to the transaction value, **trading small quantities of electricity** (microtransactions) using blockchain technology could become economically viable. In the field of electromobility, the blockchain could also create added value through **billing procedures for charging e-vehicles**. For example, the car could automatically log on to the charging station during the parking process, which would automatically bill the current output after the charging process using blockchain technology. However, the blockchain must first prevail over other advanced payment models (Hasse et al. 2016, p. 18). From a technical system perspective, blockchain technology can **simplify** many **processes** such as switching electricity providers by reducing the need for intermediaries (disintermediation). An **automated** payment of charges, levies, fees or remunerations by means of blockchain is also feasible. This would ultimately make it possible to reduce costly documentation processes for all parties involved (Bundesverband der Energie- und Wasserwirtschaft 2017, p. 30). In addition, the blockchain offers a solution for the electricity market, which is becoming ever smaller, more decentralised and shorter term. As more and more individual components of the energy supply have to be combined with each other, this circumstance pushes some previous technologies to their limits and opens up opportunities for decentralised, directly networked structures such as blockchain (Preiß, 2018b). The blockchain would make data exchange more efficient and load and generation forecasts more accurate. This would make it possible to realize a **real-time energy economy** in which the behavior of numerous devices could be coordinated based on market and network signals. In order to achieve this, it is necessary to carry out the microtransactions safely and efficiently and to make them comprehensible. The blockchain technology could make a central contribution at this point (Bundesverband der Energie- und Wasserwirtschaft 2017, p. 31). Real-time trading can open up **flexibility options** such as load management and optimal integration of storage systems (Maier 2018, p. 8ff.). From a social perspective, the greater **transparency** for consumers is particularly noteworthy. It is possible for consumers to understand exactly where the electricity they buy comes from. Due to the direct transaction between energy providers and energy consumers, a precise indication of the contractual partner and thus also the precise determination of the origin of the electricity is possible. The transparency also includes the entire transaction history (energy consumed and payments made), which is stored in the blockchain and thus provides an unprecedented degree of overview. As already mentioned in 3.6, the prosumers cannot currently participate economically in energy trading due to their low performance. Through lower transaction costs and simpler billing, the blockchain also offers small suppliers or energy consumers an opportunity to **participate in the energy market** or to act as buyers and suppliers in the market. For example, the operator of a private solar plant can sell the produced electricity more easily to neighbours or feed it into the grid and thus reach the threshold of profitability more quickly. This creates an incentive for an increase in the number of prosumers and thus leads to a further expansion of renewable energies due to the simplified marketing possibilities for decentralised energy producers (Hasse et al. 2016, p. 34f.). In addition, consumers would have the option, on the basis of real-time data, of adapting their electricity purchases to the supply at short notice and changing supplier or tariff at short intervals depending on the price and origin of the electricity, which would provide **flexibility** (Maier 2018, p. 8ff.).

Challenges

On the way to establishing the blockchain in the energy industry, however, challenges are also pointed out. In addition to economic, technical and social aspects, legal aspects also play an important role. From an economic point of view it is questionable whether the **cost-benefit ratio** of the technology is at all positive in view of the high investment requirements in infrastructure, sensor technology, measurement and control technology. In addition, there would be a risk of a **monopoly forming** if newly collected data converged with a few companies, suppliers or developers. From a technical point of view, particular attention should be paid to the **high energy consumption**, which can be traced back to the POW consensus mechanism primarily used today, which is associated with a high and constantly increasing computing effort. At this point it should be mentioned that energy consumption can be drastically reduced, for example, by using the POS consensus mechanism. The high computing effort also leads to a low **transaction speed**. In the case of Bitcoin, only seven transactions per second are possible, while VISA allows 4,000 transactions per second (Bogensperger et al. 2018a, p. 38). Another challenge can be traced back to **illegal activities**. In addition to the risk of hacker attacks, the technology is also susceptible to misuse by organized crime (e.g. money laundering). **Technical compatibility** should also not be neglected. After all, smart meters are the data sources for decentralised energy markets that are driven by smart contracts. Their interfaces must be digital and open, so that a simple connection with blockchain applications is possible (Glatz 2018, p. 31). As can be seen from the Federal Government's answer of 09.11.2018 to the *"Kleine Anfrage der Abgeordneten Ingrid Nestle, Tabea Rößner, Dieter Janecek, weiterer Abgeordnete und der Fraktion BÜNDNIS 90/DIE GRÜNEN - Drucksache 19/4823"* regarding the safeguarding of the digitisation of the energy system transformation, the Federal Government is of the opinion that the technical minimum requirements for iMSys according to the MsbG basically also support applications of blockchain technology (Federal Government 2018b, p. 7). As an FfE study shows, however, the SMGWs planned in the Smart Meter Rollout have so far not been directly compatible with the blockchain, but only via detours (external market participants) (Bogensperger et al. 2018b, p. 191). Of particular importance for these applications are the current developments of GDEW, which above all regulates the mandatory installation of intelligent measuring systems for reading and transmitting the energy demand on the part of consumers and energy producers (Hasse et al. 2016, p. 19). From a social point of view, there is a fundamental risk of **complete data loss** if access to the access data is lost. Due to the lack of a central authority to help with such problems, values, contracts and evidence would be irreversibly lost. Moreover, it is not yet clear how different blockchains could cooperate with each other (**interoperability**). After all, the user benefit depends strongly on the number of participants in the networks. The more blockchains that can interact, the greater the potential user benefit. The central challenge is that assets on a blockchain cannot be transferred directly to another blockchain, but must be exchanged via an intermediary (Bundesverband der Energie- und Wasserwirtschaft 2017, p. 47). According to Voshmgir (2016), the greatest challenges in the energy sector do not lie in the technical area, but in the regulatory area. Small players in particular are confronted with high transaction costs due to administrative and regulatory hurdles that prevent them from entering the market. A blockchain-based energy supply and transaction system would change or redistribute many market roles. It is still unclear how the **tasks and roles of market participants** are defined and why regulatory regulation is required (Sieverding and Schneidewindt 2016, p. 4). For example, an electricity supplier would need to be licensed as an energy supply company and as such would also require a business registration. Every energy consumer in a P2P network, for example, would automatically become the balancing group manager. He would therefore have to fulfil the corresponding requirements with regard to security deposit and risk management and apply for approval as an MSB. In addition, the regulations of the BNA must be technically mapped and the timetable reported to the TSOs (Geiselhardt, 2017). In addition, **data protection** is of crucial importance. After all the blockchain theoretically creates a consumption and transaction history of each participant that is stored indefinitely. Although this cannot be directly assigned by name, it can be stored in many places and thus read out by all participants. The data protection basic regulation (DSGVO) demands a right to deletion of data and to be "forgotten". In the blockchain, however, this was practically impossible or only associated with a great deal of effort (Sieverding and Schneidewindt 2016, p. 4). Also **liability rules and consumer rights** for e.g. reverse transactions have not yet been clarified.

5.3   Use Cases of Blockchain Technology in the Energy Sector

The use cases of blockchain technology in the energy sector are manifold. This diversity is particularly evident in a study by FfE, which includes a total of 91 potential use cases of blockchain technology in the energy sector that were identified within the framework of the B10X project together with Innogy, SMA, SWA, thüga, TransnetBW, VBEW, Verbund and VKW (FfE, 2018). As part of a current study by dena in cooperation with numerous industry partners on the subject of „Blockchain in der integrierten Energiewende", which was published in Berlin on 26 February 2019, potential use cases of blockchain technology in the energy industry were examined with regard to technological maturity, competitive situation with other digital technologies, business and economic benefits, strategic added value and regulatory environment (dena, 2019b). The selected use cases were assigned to five overarching application groups, which are shown in Fig. 5.6 below.

Group	Assetmanagement	Data management	Market communication	Trading	Finance & Tokenization
Use Cases	• Bottleneck Management in Electricity Distribution Networks (E-Mobility) • Energy Services for Buildings & Industrial Processes (Maintenance)	• Registration of investments in the market master data register • Certification of guarantees of origin	• Billing of fees and charges • Termination and change of supplier	• Tenant electricity • Trade and allocation of network capacity • P2P trading between customers of an electricity supplier • Off-exchange Wholesale	• Shared investments with external tenant electricity

Fig. 5.6: Application groups and use cases. Source: Own presentation based on Richard et al. (2019): p. 16.

As a result of numerous assessments by experts from the energy industry, several studies and two surveys, there is agreement that P2P trading in particular can be considered as a potential use case based on blockchain technology. A showcase project is the *"Brooklyn Microgrid"* of New York-based LO3 Energy, which in April 2016 realized the first direct sale of decentralized energy to neighbors via a blockchain system. Such applications network prosumers with each other or connect energy providers directly with consumers, making blockchain technology a trailblazer for further decentralization of energy systems (Hasse et al. 2016, p. 15). In the context of GDEW, so-called "prosumer communities" are of particular interest, in which prosumers sell their locally generated electricity directly to other end customers and thus intermediaries such as the energy supplier or trader are no longer needed.

6 Concept for P2P Electricity Trading via Blockchain

A potential assessment has shown that there are different possible combinations for P2P electricity trading. The decentralised P2P trade, which enables a new form of interaction, is particularly promising. Thanks to the decentralised transaction and energy supply system, intermediaries such as electricity suppliers and electricity traders/exchanges could be eliminated. However, an analysis of the regulatory framework shows that such a concept for prosumers is only possible under difficult energy and bureaucratic conditions. This could be remedied by a service model in which all regulatory obligations are assumed by the platform user as a service provider. A business model, which was developed independently, comprises an infrastructure provided for this purpose, a detailed description of the individual process steps within the framework of a business process map as well as a potential design of the blockchain.

6.1 Introduction

P2P electricity trading is a concept which is based on the principles of a smart market and provides direct access to the electricity market without central instances such as energy suppliers through interacting market players. Basically, there are different combination possibilities between the market participants acting together. The tasks of the exchange and OTC trading in the B2B area can be mapped by the technology. In addition, exchanges and the role of suppliers in the form of energy supply companies and aggregators can be substituted by trading directly between prosumers in a P2P model using blockchains. The latter possibility is a new form of interaction. For this reason, the following considerations focus on the third option. While the P2P traded energy quantities still flow through the public distribution network, the marketing and electricity procurement according to this concept is no longer carried out via a conventional energy supplier, but via a P2P network between the end customers. The decentralised transactions, which do not require a central trader or distributor, could drastically reduce transaction costs, speed up trading and make it more secure. Since so far hardly any empirical values on the concrete implementation of P2P trading have been published, a potential assessment should provide information about the added value the blockchain offers in this case, how the market potential is to be assessed and which regulatory obstacles exist. On the basis of the insights gained, a concrete business model for electricity trading based on the blockchain will be developed on the basis of our own theoretical considerations.

6.2 Problem Formulation

The energy system transformation in Germany aims at the transition from conventional energy sources to renewable energies. In 2017, one third of the total gross electricity generation in Germany could already be generated by renewable energies. Due to the increasing decentralization of energy supply, power generation is no longer in the hands of municipal utilities and large corporations, but increasingly also of private individuals, municipalities, companies and citizen energy projects. As a result, almost one third (31.5 %) of the installed capacity was provided by private individuals for regenerative electricity production. Prosumers in particular therefore have an important role to play; they not only consume their own electricity but also sell it to third parties and are therefore both producers and consumers of energy. *"Prosumers are becoming increasingly important for the energy revolution"* (BMWi, 2016). Although prosumers make a significant contribution to energy system transformation, they are currently not represented in the actual market, since the marketing of electricity continues to be in the hands of aggregators such as the grid operator or the direct marketer. The processing of all transactions via the traditional energy suppliers represents a significant barrier to market entry for small producers with low services because the transaction costs are too high in relation to their transaction value due to administrative and regulatory hurdles (Voshmgir 2016, p. 24). Since prosumers cannot therefore participate economically in electricity trading and fail due to legal and bureaucratic hurdles, the future energy market is threatened with the loss of a decisive player (Glatz 2018, p. 7f.). While the actors are already decentralised, the market and trade remain stuck in centralised structures (Bergmann, 2017). The challenge of energy system transformation to date has therefore been the decentralisation of electricity trading. According to Voshmgir, a large part of the value added in energy trading could take place locally and regionally via P2P trading and thus initiate a further phase of the energy system transformation in Germany (Voshmgir 2016, p. 24). New business models must be created to enable smaller market participants to participate in the market with low transaction costs. With a view to new technological possibilities, blockchain technology in particular could prove to be a particularly effective link for processing transactions securely, cost-effectively and automatically, thus strengthening the role of prosumers. P2P trading is regarded as a use case that can bring about promising changes in the energy sector on the basis of blockchain technology. As a result of many assessments by experts from the energy industry, the potential for P2P trading can basically be regarded as large. The forms of P2P trading are diverse and (theoretically) allow the disintermediation of exchanges, brokers or even energy suppliers (Bogensperger et al. 2018b, p. 35). While the first pilot projects in the USA, such as the *"Brooklyn Microgrid"*, are testing the direct sale of locally generated solar energy to neighbors via a blockchain for the first time, such applications are inconceivable in Germany to date due to a lack of practical experience and technical and regulatory challenges (Sieverding and Schneidewindt 2016, p. 2f.). For example, it has been clarified at best for tenant electricity models and own consumption which allocations are due at which point in the course of the energy system transformation (EEG allocation, grid fees, etc.). For P2P energy trading, this question has not yet been clarified (Voshmgir 2016, p. 24). Another unresolved question is whether and how the plants will continue to be operated after expiry of the subsidy under the EEG, which guarantees a fixed feed-in tariff for generally 20 years. In addition to the consumption of the electricity by the customer itself, there is still a right to connection to the grid and purchase of the electricity generated after the subsidy has expired. However, within the framework of "other direct marketing" pursuant to § 21a EEG 2017, a contract must be concluded with an energy supply company, a municipal utility or another electricity trader who purchases the electricity. In this case, it merely offers the operator of the PV system the opportunity to collect the proceeds to be generated on the electricity market (Wagenblass, 2018). However, it is questionable whether the expected revenues or the savings to be achieved will cover the costs of operating and maintaining the old system. With blockchain-based P2P trading, the operation of a PV system could continue to be economically viable if the electricity generated were to be directly transmitted between the consumer and the producer (Linden, 2018).

6.3 Objectives and Questions

The overriding aim of the concept is to find a solution for how prosumers can participate economically in electricity trading and thus continue to provide significant support for the transformation of energy systems. To this end, the potential of P2P trading in Germany, in which electricity in particular is traded directly between producer and end consumer by means of blockchain technology without detour via municipal utilities and electricity providers, will first be investigated and evaluated. On the basis of the potential assessment, a concrete business model is to be developed which will show the necessary infrastructure, a detailed process description and design options for the blockchain. Essentially, the following questions will be answered:

- What is the potential of P2P trading based on the blockchain?
    - Is a blockchain basically an option for P2P trading?
    - Can intermediaries be neglected?
    - What are the regulatory challenges?
    - What is the market potential?
    - What are the business opportunities?
- What does a potential business model look like?
    - What infrastructure is required?
    - What are the interdependencies between the various players?
    - What do concrete process steps look like?
- How should the blockchain be designed from a technical point of view?
    - Type of blockchain

- Consensus mechanism
- Participants
- Authorization (Public Key/Private Key)
- Smart-Contract
- Block structure
- Platform for implementation

6.4 Evaluation of potential

The blockchain technology does not offer advantages over existing technologies for every possible use case, emphasizes the BCI-E. *"Rather, it is advisable to take a close look at and evaluate the respective use case"* (Preiß, 2018c). For this reason, a potential assessment is carried out for P2P trading.

Assessment of Technical Suitability

The technical suitability of the blockchain technology for P2P electricity trading has already been demonstrated by projects such as the Brooklyn Microgrid and will therefore not be discussed further.

Visualization of the Process with Simplified e3-value Method

Traditional electricity trading comprises a number of participants with different market roles. Fig. 6.1 shows the most important players, whose market roles and interrelationships are described in more detail below.

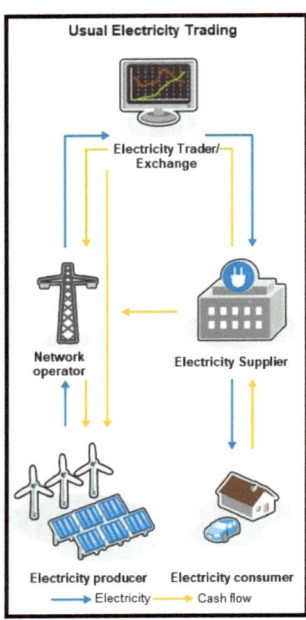

Fig. 6.1: Functional diagram of classical electricity trading. Source: Maier (2018): p. 9.

First, electricity is produced by the **electricity producer**. In the regenerative sector, electricity production can be based primarily on wind, hydropower, photovoltaics, biomass or geothermal energy, and in the conventional sector it can take place primarily via power plants using fossil fuels such as lignite or hard coal, natural gas, oil or uranium. In the course of the energy transition, the prosumer is increasingly establishing itself as a market participant, combining the role of the electricity consumer and that of the electricity producer in one person. The electricity produced is passed on to the **network operator**, who is responsible for the infrastructure and the transport routes. In addition to ensuring grid stability, the grid operator is also responsible for ensuring that the electricity is transmitted from the high-voltage grid via the medium-voltage grid to the low-voltage grid. In an intermediary function, the **electricity supplier** is responsible, among other things, for procuring the required quantities of electricity. As a rule, electricity is procured via organised trading centres such as the Leipzig **Electricity Exchange** (European Energy Exchange or EEX for short). The forecasted electricity consumption is reported daily by the electricity supplier to the grid operator. After all, the electricity supplier supplies all **electricity consumers** such as households or companies with electricity. Other tasks of the electricity supplier include ensuring that all processes run smoothly in the background and billing all levies and allocations in addition to the energy costs. The procurement strategy and the electricity supplier's own fees influence the electricity costs for the electricity consumer (cf. EHA, 2018).

Comparison of the Use Case with and without Blockchain

If one compares the classical process of electricity trading with the process using blockchains, it becomes clear that the role of the electricity supplier and that of the electricity trader/exchange are eliminated (Fig. 6.2).

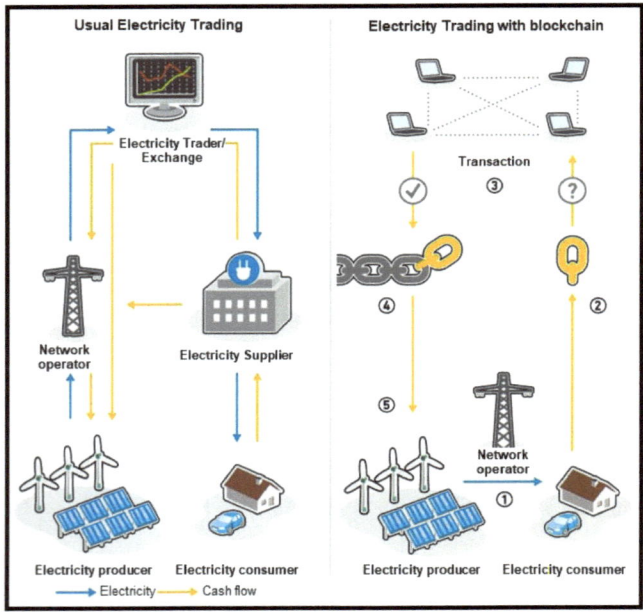

Fig. 6.2: Comparison of Electricity Trade with and without Blockchain. Source: Maier (2018): p. 9.

This is due in particular to the value proposition of blockchain technology in enabling direct interactions between peers. The result is a decentrally controlled transaction and energy delivery system that can be described using the following five steps (Maier 2018, p. 9):

(1) The generator supplies **electricity** directly to an electricity **consumer** in its vicinity.
(2) Each transaction is combined with other transactions to form a data block.
(3) The transaction is checked and confirmed in the network on decentralized computers.
(4) The block is appended to the existing data chain (blockchain).
(5) The electricity producer receives the payment.

Identification of regulatory Barriers

For the identification of the regulatory obstacles, the regulatory report of the auditing company Deloitte can be consulted, which was used in the dena study on "Blockchain in der integrierten Energiewende". The following main regulatory obstacles were identified:

- **§ 3 No.18 EnWG**: As a supplier, the prosumer is also an energy supply company within the meaning of the EnWG, which gives rise to further obligations.

- **§ 5 EnWG**: Obligation for suppliers in the (permanent) supply of residential customers (personnel, technical and economic capacity) to notify the regulatory authority.

- **§ Section 41 EnWG**: Minimum requirements for the contractual design of energy supply contracts.

- **StromNZV § 1 sentence 1** regulates the condition for feeding electricity into the grid. The contractual form of useful access must comply with the requirements of **§§ 23 et seq. EnWG** of the Electricity Network Access Ordinance. EnWG.

- **§ 4 StromNZV**: Designation of a BKV which reports load forecasts to grid operators.

An analysis of the regulatory framework shows that direct P2P trading on a blockchain basis is possible for prosumers, but is hampered by many energy and other bureaucratic hurdles (Richard et al. 2019, p. 208). Although from the point of view of the prosumer it would be possible to fulfil these obligations, their basic classification as EVU should already have a considerable deterrent effect. This could result in potential participants of a direct P2P network ultimately refraining from such participation. To counteract this, the tasks and obligations can be assumed by a service provider such as the energy supplier or an electricity supplier, which is therefore responsible for balancing group management, the procurement of replacement and surplus quantities, forecasts and bureaucratic obligations (notification of the prosumer as an energy supply company, payment of the EEG levy, electricity tax obligations, etc.), while the participants within the balancing group can act freely among themselves. This would allow ex-post P2P trading on the blockchain (Bogensperger et al. 2018b, p. 129). Since the regulatory barriers to direct P2P trading are currently considered too high, the focus in further consideration is on the legally compliant solution approach based on a service model.

Visualization of the Blockchain Solution with Business Model Canvas

For the description of P2P trade in the context of a service model, the use of the Platform Business Model Canvas is appropriate due to different actors and roles. Essentially, the following four actors could be identified, which will be explained in more detail in the following sections:

- Platform provider
- Platform users
- Data provider

- Data users

The **platform provider** is responsible for the development and advancement of the platform, in particular the provision of the blockchain technology. In addition, the platform provider markets the platform to companies (= platform users), such as IT companies, public utility networks or associations or companies with blockchain competencies, who use the common database structure for their individual value proposition. In return, the provider receives a commission for e.g. maintenance and support, data preparation and the development of smart contracts (Bogensperger et al. 2018b, p. 131).

After the **platform user**, such as municipal utilities, electricity suppliers and EVUs with direct customer ties, has paid a commission to the platform provider, he receives read and write access for P2P trading on the trading platform. This includes a large number of existing smart contracts and a functioning blockchain solution, which enables him to connect his end customers' facilities (producers, consumers, prosumers) to the platform and thus offer them an individual product. The main tasks of the platform user are primarily in the energy industry activities such as balancing group management, supplier change and regulatory notification obligations, which he also assumes as a service provider for the data suppliers (Bogensperger et al. 2018b, p. 131).

The **data supplier** is primarily the end customer of the platform user, who is a generation or consumption system or end consumer or prosumer. The platform gives the data suppliers a new opportunity to purchase energy quantities (consumer) or sell them to other end consumers (prosumer). In return for providing data on their production or consumption, the data suppliers are given the opportunity to participate in P2P trading. In addition to greater co-determination and a free choice of supplier, they also receive the energy management handling of all regulatory processes by the platform user. However, participation in the system requires secure hardware for digitizing generation and consumption data (iMSys). Participation in the consensus procedure is not planned (Bogensperger et al. 2018b, p. 132).

The **data user** only has read access to the blockchain to evaluate the data in the distributed database. In addition to portfolio management, forecasting or benchmarking of plants, the data can also include a database for regional investment projects or crowdfunding by EE, which can be used for a variety of data-based business models. The latter is possible because the preferences of the final consumer are stored regionally in the blockchain and can be derived from the locally available producers which demand cannot be met (Bogensperger et al. 2018b, p. 132).

Assessment of the Market Potential

The potential assessment shows that blockchain technology is (theoretically) able to substitute classic market roles such as electricity supplier or electricity trader/exchange and instead enable electricity trading directly between producers and consumers. The analysis of the theoretical potential indicates that the potential is very high, especially in the area of C2C, and that blockchain technology can offer real added value here due to its value propositions such as decentralisation, actor diversity, manipulation security and the possibility of carrying out microtransactions. However, very far-reaching regulatory obstacles make it difficult to use blockchain technology in direct P2P trading in the short term, so that there would probably be few interested parties to participate in such an organised electricity trade. If, however, a platform is offered as part of a service model in which all regulatory obligations are assumed by the platform user as a service provider, a blockchain solution can be created that enables electricity trading between prosumers and consumers in a legally compliant manner and at the same time makes use of the advantages of blockchain technology. The following section describes a business model based on such a service model.

6.5 Business Model

The potential assessment shows that a fully decentralised, blockchain-based P2P trading platform is difficult to reconcile with the existing legal framework. Nevertheless, in order to make use of the advantages of blockchain technology on the one hand and to bundle the associated legal and regulatory risks on the other, a business model based on a service is to be developed. P2P electricity trading will be implemented via its own online trading platform, whose operator is a service provider such as an electricity supplier. The service provider alone is responsible for the energy industry and regulatory obligations as well as the balancing group responsibility, so that all essential areas of responsibility of the P2P participants are transferred to it. Using this trading platform, prosumers and consumers could trade electrical energy automatically, securely and cost-effectively with each other and through the use of smart contracts. The blockchain technology ensures that all information on production and its marketing is documented for each plant. In this way, each kWh produced can be clearly assigned to a prosumer, which guarantees proof of origin at all times. While prosumers can stop and sell their offers, consumers can choose purchase conditions such as price or the composition of their electricity supply via the platform or change suppliers. As part of a community approach, electricity suppliers with PV systems and home storage facilities can exchange electricity nationally with customers from the same service provider via the trading platform. Although this business model would not achieve the independence of end customers from energy supply companies or similar service providers, higher sales revenues or lower electricity costs, improved transparency of the origin of electricity, a local electricity market or improved integration of renewables could still be achieved, depending on the specific design of the P2P network. Before a business model can be designed, the **assumptions and prerequisites** must first be made in an initial step. In a further step, the necessary **infrastructure** is developed and all relevant, value-adding **process steps, functions and interactions** are described.

Assumptions and conditions

Unlike other blockchain applications, which operate exclusively within a digital framework, P2P electricity trading involves the physical delivery of the quantities of electricity traded. Therefore, it is necessary that all network participants are not only digitally connected over the Internet, but also **physically connected to each other**. Since each household is usually already connected to the local distribution network, it is obvious that this connection will also continue to be used for feed-in and supply within the framework of P2P trading. Ideally, however, households should be connected to each other in a **smart grid**, which contains means of communication and intelligent components such as **smart meters** and **SMGW**. While a smart meter is required to accurately capture actual energy consumption and usage times, the SMGW enables integration into a communications network, serving as an interface between physical power flows and trading transactions through the blockchain. Together they form the iMSys, which is a mandatory prerequisite for P2P trading using blockchain technology, as this is the only way to accurately record consumption and, if necessary, the quantities of electricity fed into the system. It is assumed that the iMSys is compatible with blockchain technology. This is the only way to ensure that the data or information collected by the iMSys is transferred directly to the blockchain or processed by linked smart contracts. Before a prosumer can offer his electricity for sale, it must first be generated. It is assumed that the prosumer has installed a **PV system** on the roof of his house. Since one cannot necessarily assume that the quantity of electricity generated by the prosumer will be taken directly by a consumer, both the prosumer and the consumer are equipped with an **electricity storage device**. This is intended to create a situation in which the prosumer can react immediately to a consumer's demand at any time. In order for trade to take place between a prosumer and a consumer, it is imperative that the respective purchase conditions of the consumer match the available capacity or supply of the prosumer.

Infrastructure

The infrastructure serves to illustrate all essential components of the business model and their relationships to each other. The central component is a **trading platform**, which resembles a local marketplace and is operated by a service provider. The main players are **prosumers** and **consumers**. At the electrotechnical level, the prosumer is physically connected to the consumer via the local distribution grid, which enables the energy generated to be transported to the end consumer. While the prosumer and consumer are each equipped with an iMSys, which comprises both a **smart meter** and a **SMGW**, the prosumer also has a **PV system** installed on the roof of the house and used to generate electricity from renewable sources. With the help of the iMSys, generation and consumption data can be measured at each network participant in defined time intervals and continuously communicated via the SMGW, thus forming the interface between the electrotechnical and information technology levels. This decisive transition is necessary in order to transfer content from the real world to the blockchain. For this purpose, the prosumer is assigned an address in the blockchain ledger, which in principle resembles a wallet. This so-called **EC-Wallet** contains information about his person (personal data), e.g. the serial number of the PV system as well as an image of the power storage level. By linking the personal data and the serial number of the PV system, a clear link can be established between the power generation and its origin (proof of origin). The image of the electricity storage status represents the currently available storage capacity and thus reflects the account balance at the IT level. Since the storage capacity is usually stated in kWh and is therefore unsuitable for data processing on a blockchain, one unit of electricity capacity (1 kWh) is equated with an EC token (1 kWh = 1 EC token). For example, the balance is synchronized every minute and forms the basis for creating an offer on the trading platform. After all, a certain amount of electricity can only be offered if it is actually available. The actual value transfer at the monetary level takes place via so-called EH tokens, which assign a defined value to their owner within the trading platform and can be exchanged for Fiat money. In principle, every prosumer and consumer has their own **EH wallet**. The operator of the trading platform also uses an EH Wallet which is implemented in perspective between the EH Wallet of the prosumer and that of the consumer and serves as a trust account. In order to prevent fraudulent acts, the EH tokens of the respective purchaser are temporarily stored on the escrow account and only credited to the respective seller once it has been proven by means of a proof-of-delivery that the agreed quantity of electricity has actually arrived at the purchaser or his electricity storage facility. On the blockchain, all actions from the sale, trading and purchase through to the electricity transfer are stored manipulation-proof and can be tracked transparently at any time. Fig. 6.3 shows the infrastructure for electricity trading between a prosumer and a consumer at the electrical, IT and monetary levels as a simplified e3-value model.

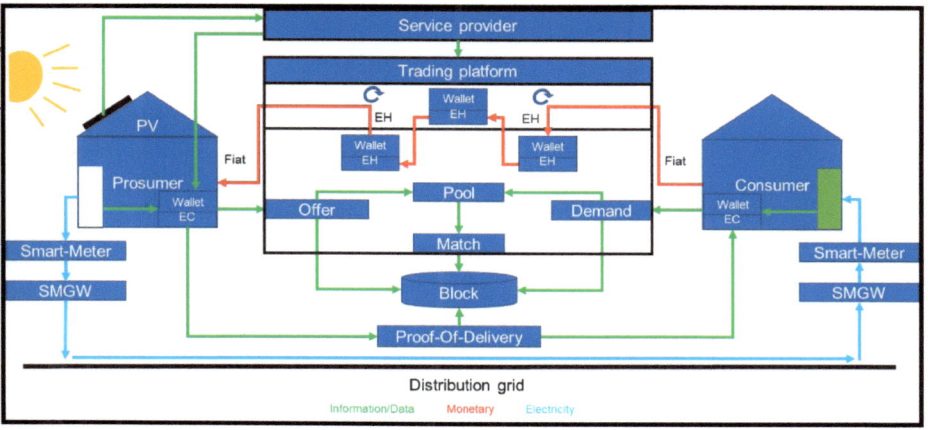

Fig. 6.3: Infrastructure as a Simplified e3-value Model for Electricity Trading between a Prosumer and a Consumer. Source: Own Representation.

Not to be neglected are Smart-Contracts, which are used for the automated interaction within the blockchain, but which are not shown in the infrastructure for reasons of clarity.

Process Steps, Functions and Interactions

In order to be able to describe and illustrate all necessary process steps, functions and interactions of the involved attributes in detail, a business process map was modeled especially for this use case (see appendix). In principle, the entire business process can be subdivided into eight consecutive steps:

- Authentication
- Electricity generation
- Tokenization
- Selling
- Electricity trading
- Buying
- Electricity transfer
- Completion

Authentication shall be regarded as a preparatory measure taken by the operator of the trading platform to ensure that all prosumers involved are clearly identifiable and that their consumption and production data can be tracked and allocated. In order to ensure this, prosumers must first register with the responsible service provider, who links the personal data of the prosumer with the data of his PV system (e.g. serial number) and assigns it to a personal EC-Wallet. This makes it possible to establish a clear link between the quantity of electricity generated and the respective prosumer at any time, thus enabling proof of origin to be provided.

Electricity generation is always generated with a PV system. The generated electricity can then be temporarily stored by means of an electricity storage system. With the help of the smart meter, the generation data and electricity consumption are recorded and transmitted to the SMGW. In order to document the data transparently and inseparably on the blockchain in a further step, each unit of the quantity of electricity generated (per kWh) is marked accordingly (proof of origin). At the same time, the labelling forms the basis for the subsequent tokenisation.

Tokenization is required to connect real objects from the real world to the blockchain. In this case, the quantity of electricity generated, which is determined by the smart meter and stored by the SMGW, is to be assigned to an equivalent quantity of a digital currency (e.g. 1 kWh = 1 EC) in order to map ownership in terms of electricity capacity and guarantee guarantees of origin. In order to be able to offer or sell a certain quantity of electricity on the trading platform, a corresponding monetary equivalent (electricity price) must be defined. For this purpose, EH tokens are used which are purchased by consumers for Fiat money and act as a means of payment within the trading platform. The value of an EH token can be defined arbitrarily, but a direct coupling to a fixed value (e.g. 1 EC = 1 EH = 0.2 €) is advisable for reasons of value stabilisation. In the following, two different variants with regard to the equivalent value are explained. The first variant (**static value**) assigns a fixed equivalent value to a token unit (1 EH). This is intended to minimize the volatility of the price and avoid speculative transactions.It should be noted here that the equivalent value of the EH token is the decisive factor in pricing. In order to give customers or consumers an incentive to use the trading platform, the value of an EH token should be below the average electricity price (< 29.44 ct/kWh). From the point of view of the seller or prosumer, the price they receive per kWh should be more than the normal market feed-in tariff (> 11 ct/kWh) (Fraunhofer 2019, p. 10). However, this requires cost savings through, for example, the neglect of conventional intermediaries. In order to determine the possible cost savings, a cost-benefit analysis is recommended. First, the costs incurred by the intermediary should be determined. The next step is to compare the costs incurred with the cost savings and to convert them into the share of the electricity price, from which the actual equivalent value can be determined. The second variant (**dynamic value**) was derived from the Bitcoin system, where the price of a unit (1 BTC) is determined by the market. Due to the limitation of the maximum Bitcoins available on the market, this is a finite supply (max. 21 million BTC), which means that increased demand inevitably results in a rising price. At the same time, Bitcoin's rising price increases the attractiveness for miners to participate in prospecting for the digital currency. In order to participate in mining, investments in mining equipment (miners) are required, which lead to a higher hash rate as performance increases. The record high of the Bitcoin price in 2017 led to a higher participation in mining and thus to a strong increase of the hash rate in the entire network. If this effect is transferred to the energy industry, the value of an EC token could be determined as a function of the installed capacity in the network. For example, the value of the EH token could increase if the installed capacity of RE systems within the network increases. The price would be directly driven by the expansion of EE plants. This could lead to an effect similar to that of Bitcoin, where entire mining farms were built, so that renewable energy plants such as wind farms or PV plants are built in order to ultimately obtain EC tokens. The expansion of renewable energy plants could make a positive contribution to energy system transformation. However, if the equivalent value is left to the relationship between supply and demand, speculative transactions, among other things, could lead to strong price fluctuations. In order to achieve a similar effect, the maximum number of EC tokens would inevitably have to be determined. However, this would not be compatible with the principle of the concept, since the EC tokens are assigned per kWh generated. Since this is exclusively electricity generated by EE and therefore (theoretically) infinite, the number of EC tokens must also be unlimited. Since the implementation of this variant is difficult and therefore presents certain hurdles, only the first variant will be considered in the further consideration of this work.

If a prosumer decides to **sell** the electricity he generates, he creates an offer based on its available EC tokens. In order to ensure that the prosumer actually has the amount of electricity offered, the EC-Wallet synchronizes itself in almost real time with the iMSys, which, among other things, transmits data on the current memory status. If sufficient storage capacity is available, the offer is considered valid and is stored on the blockchain with a so-called "Sell TAG". This TAG contains information such as the time at which the offer was created, the quantity of electricity offered and the sales price. At this point, it must be determined whether the quantity of electricity offered by the prosumer can also be consumed at any time after the offer has been prepared or whether access to the quantity of electricity offered is denied during the offer. Since it would be technically and also legally more problematic to deny access to the quantity of electricity generated by the prosumer himself, the prosumer has in principle the possibility to consume the quantity of electricity offered himself. However, if the available quantity of electricity falls below the offered quantity in the meantime, the offer is deactivated until sufficient electricity capacity has been proven. If an offer is valid, it is available with further offers in a pool for electricity trading.

In order to participate in **electricity trading** as a consumer, you must first register on the trading platform. Since electricity trading is based on a blockchain or a digital cash book, Fiat money is not suitable as a tradable currency. For this reason, every consumer must first exchange Fiat Money for so-called EH Tokens at the operator of the trading platform. The operator can charge a corresponding transaction fee for this service. Upon registration, the consumer receives his or her own EH wallet on which the exchanged EH tokens are credited and stored. Next, the consumer defines his purchase conditions, such as the origin of the electricity or the price. The demand is then compared with the offers from the pool in terms of matching. Once a suitable offer has been found, a Smart-Contract is used to check whether there are sufficient EH tokens on the consumer's wallet. If sufficient EH tokens have been proven, it is also checked whether the quantity of electricity offered has not already been sold or consumed (double spend). If sufficient EH tokens have been proven by the consumer and a double sale has been excluded (match), the trading result is stored on the blockchain with a so-called "Trade TAG". This TAG contains, for example, information on the buyer, seller and a trading protocol, which includes the agreed quantity of electricity and the agreed price.

(**Buying**) After all essential criteria for successful trading have been met, the agreed, current number of EH Tokens of the consumer is automatically transferred to the platform's own EH Wallet (escrow account) using a smart contract and stored there in the meantime. At the same time, a so-called "Buy TAG" is created on the blockchain, which contains the time of demand, the quantity of electricity demanded and the purchase price.

(**Electricity transfer**) As soon as the corresponding number of EH tokens is completely on the trust account of the operator, the agreed quantity of electricity is transferred from the electricity storage of the prosumer to the electricity storage of the consumer via the distribution network. Once the agreed amount of electricity has been fully transferred, another "Electricity TAG" is created on the blockchain, which contains information about the meter readings of the accumulators before the transfer, the amount of electricity being transferred, and the meter readings of the accumulators after the transfer.

(**Completion**) Once it has been ensured that the agreed quantity of electricity has been transferred, the EH Tokens are transferred by means of Smart-Contract from the operator's escrow account into the possession of the prosumer, who can then either use the EH Tokens himself to purchase electricity or exchange them for Fiat money. If the prosumer decides to exchange the EH tokens for fiat money, a corresponding transaction fee is charged to the operator. At the same time, the EH tokens are automatically transferred from the prosumer's EH wallet to the consumer's EH wallet via Smart-Contract, which guarantees proof-of-delivery. Finally, all TAGs (Sell, Trade, Buy, and Electricity) are combined in a single transaction and stored on the blockchain, enabling transparent, traceable and tamper-proof electricity trading.

Design of the Blockchain

After the business model has been described, this section contains considerations on the concrete design of the blockchain. The following characteristics are taken into account:

- Type of blockchain
- Consensus mechanism
- Participants
- Legitimacy
- Smart contract
- Block design
- Platform for implementation

As already described in Section 4.3, a distinction is essentially made between public, private and consortium blockchains. The choice of **blockchain type** depends on the requirements of the respective use case. In P2P trading, high transaction speeds are of particular importance in addition to energy-efficient validation of transactions and compliance with data protection. Since the energy-intensive POW consensus mechanism is generally used for public blockchains, data protection is restricted and transaction speed is limited, this type of blockchain can be categorically ruled out for P2P trading. Compared to a public blockchain, a private blockchain can be operated more efficiently and faster, as consensus mechanisms such as POS or POA are generally used, which have low energy consumption. While in a private blockchain the validating nodes are only operated by one legal entity, in a consortium blockchain this task is distributed among a consortium of several organizations in the network. Thus updates and changes can be decided by the majority principle in the consortium. This is especially advantageous in a highly regulated energy market to be able to react quickly to possible legislative changes. In addition, the encryption level requirements and the verification of new blocks can be reduced by restricting access in a consational blockchain, so that the transaction rate can increase. Overall, consortia blockchains offer the possibility of being adapted to the specific requirements of the energy market (Bundesverband der Energie- und Wasserwirtschaft 2017, p. 69). Tobias Federico, Managing Director of the management consultancy Energy Brainpool, believes that the consortium blockchain will find its place in the energy world in the future (Bollinger-Kanne, 2017). For these reasons, a consortium blockchain, especially in combination with an optimized consensus mechanism, is to be preferred for P2P trading.

In principle, all **consensus mechanisms** described can be used on the basis of a consortional blockchain. In contrast to the POW, the POA uses validators and does not use mining, which means that no high computing power is required for validating new blocks. The POA also has an advantage over the POS because users do not need a large number of tokens to qualify as validators. The POA consensus mechanism is primarily used in consortia blockchains, since validators must first be authorized in order to participate in the consensus mechanism. In the case of P2P trading, the role of the validator is the responsibility of the operator of the trading platform. If several trading platforms are operated simultaneously, the role of the validator within the consortium can be extended or transferred to the authorised operators of the other trading platforms.

The main **participants** have already been identified within the framework of the potential assessment. While the platform provider is responsible for the development and further development of the platform and provides the blockchain, the platform user forms part of the consortium as e.g. energy supply company or electricity supplier. The data suppliers include prosumers and consumers who only act as users and do not participate in the consensus procedure. The data provided by the data suppliers, such as yield and consumption information, form the basis for evaluation by the data user.

Since only the platform operator is authorized to validate new blocks, it has **read and write authorization**. As part of the consortium, the platform user also receives voting rights within the framework of the POA consensus mechanism. This allows the user to participate in the design and further development of the trading platform and the underlying blockchain. The data user only receives read rights on the consortium blockchain in order to evaluate the data in the distributed database.

A **smart contract** works according to the "if-then" principle. If this logic is applied to P2P trading, the use of a Smart-Contract can first ensure that a trade between a prosumer and consumer is only possible if the consumer's purchase condition matches the prosumer's offer. Smart-Contracts can also be used to check further criteria such as sufficient credit on the part of the consumer or the exclusion of a double sale. Once all the criteria required for trading have been met, the consumer's agreed current number of EH tokens is automatically transferred to the platform's own EH Wallet (escrow account) using a Smart-Contract and stored there in the meantime. Once the agreed amount of electricity has been successfully transferred, the EH tokens are automatically transferred from the operator's escrow account to the prosumer's possession by means of a smart contract. To ensure proof-of-delivery, another Smart-Contract is executed at the same time. This ensures that the EH tokens are transferred from the EH wallet of the prosumer to the EH wallet of the consumer. Ultimately smart-contracts combined with the characteristics of a blockchain contribute significantly to the fact that a trade between two unknown parties can take place securely and without personal trust. Thus, the prosumer can be sure that he will be paid for his electricity, while the consumer can be sure that he will receive the paid electricity delivery.

The block can basically be divided into a block header and a block body. In addition to the hash of the previous block, which is necessary for concatenating the blocks, the block header also contains a time stamp and a Merkle root hash tree, which contains the compressed hash of all transactions. The block body contains the individual transactions, each of which consists of four TAGs. These TAGs contain specific information on the sale, trade, purchase and delivery of electricity. Fig. 6.4 shows an example of all relevant block contents.

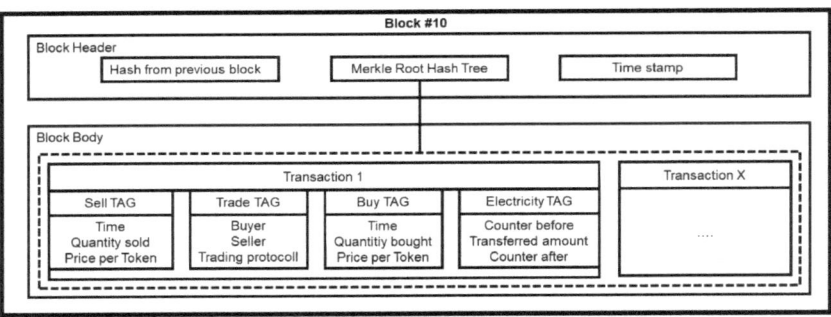

Fig. 6.4: Exemplary Block Contents. Source: Own Representation.

As a recent article on *"Blockchain technology in the energy sector: A systematic review of challenges and opportunities"* shows, most (50%) use cases in the energy sector are based on the Ethereum platform (Andoni et al. 2019, p. 157). The Ethereum blockchain basically supports smart contracts, which are also used within the framework of the concept. For the execution of Smart-Contracts, however, transaction fees are incurred. In addition, the Ethereum platform is based on the energy-intensive POW consensus mechanism. In addition, there are legal hurdles against the background of the new EU data protection regulation. *"The data in the smart contracts are written one-to-one unencrypted into the Ethereum blockchain and can then be viewed by participating third parties"*, says Merz (Kloth, 2018b). For these reasons, the Ethereum platform is unsuitable for the concept. A promising solution is offered by the EWF, which is currently working on a platform based on a consortium blockchain and the POA consensus mechanism through which energy trading will be conducted in the future. This should make the blockchain significantly faster and consume less energy. The new procedure also enables many transactions to be carried out in a short period of time, which is essential for a P2P energy trading platform, for example. According to Ewald Hesse, Vice President of the EWF, an encryption concept for smart contracts will also be used to make blockchain technology faster and more resource-efficient for the energy sector (Kloth, 2018c).

6.6 Key Findings and Recommendations for Action

For P2P electricity trading, there are different possible combinations between the market participants acting together. The decentralised P2P trade, which enables a new form of interaction, is particularly promising. Initial projects such as the "Brooklyn Microgrid" confirm the technical suitability of the blockchain for this special area. Thanks to the decentrally controlled transaction and energy supply system, which allows direct interactions between peers, intermediaries such as electricity suppliers and/or electricity traders/exchanges are eliminated. An analysis of the regulatory framework shows that such a concept for prosumers is only possible under difficult energy industry and bureaucratic conditions and can therefore be classified as unacceptable. This is remedied by a service model in which all regulatory obligations are assumed by the platform user as a service provider. This could lead to legally compliant electricity trading between prosumers and consumers, which nevertheless makes use of the advantages of blockchain technology. A business model, which was developed independently, comprises an infrastructure intended for this purpose, a detailed description of the individual process steps within the framework of a business process map as well as a potential design of the blockchain. In order to enable secure electricity trading between prosumers and consumers, a trading platform is used which is operated by a service provider (e.g. electricity supplier). From a technical point of view, some assumptions have been made. In addition to communication-capable hardware (Smart-Meter/SMGW), which is compatible with the blockchain, the possession of an electricity storage device is also a prerequisite. In order to map physical quantities such as the storage capacity (in kWh) on the blockchain, an EC token is assigned for each kWh, which also serves as proof of origin. Payment is made on the trading platform with EH tokens, which have a fixed equivalent value for reasons of value stability and are purchased by consumers using Fiat money. Prosumers who sell their generated electricity via the trading platform receive the agreed number of EH tokens credited. In principle, it is possible to purchase electricity via the platform yourself or to have the corresponding equivalent value paid out in Fiat money. The entire payment process can be automated and carried out securely using Smart-Contracts. Smart-Contracts also offer real added value for checking credit balances or excluding double sales. Due to the energy-efficient validation of transactions, data protection compliance, limited access and the resulting high transaction speeds, a consortium blockchain using the POA consensus mechanism is preferred. In order to check whether the business model also makes economic sense, a cost-benefit analysis is recommended. If the result of the analysis is an economic added value, the next step is to implement the business model as part of a proof-of-concept.

7 Conclusion and Outlook

The energy system is undergoing a process of transition worldwide and in Germany in particular, which is referred to as the "Energy Transition ". As a result of the associated increasing decentralisation, many grid customers will develop from pure consumers to prosumers in the future. They thus form the starting point and central element of the future energy system. At present, however, prosumers cannot participate economically in electricity trading because the transaction costs are too high in relation to their transaction value due to administrative and regulatory hurdles (Voshmgir 2016, p. 24). In addition to the emergence of new market players and roles, far-reaching changes in generation and grid infrastructures are also imminent (Reck, 2011). The result is a wide range of challenges, but also opportunities for the transformation of energy systems. The transformation of the energy system brings with it a complexity that can only be mastered with a high degree of automation. In particular, the cost of decentralised power generation and flexibility systems is increasing and requires a corresponding control and IT solution. The inclusion of a large number of individual systems in the low-voltage grid also requires an efficient communication and trading system between market participants (Verband der Elektrotechnik, Elektronik und Informationstechnik 2007, p. 91f.). This makes it necessary to collect and exchange large amounts of data. In addition to these aspects, there are also high data security requirements associated with the exchange and billing of electricity. The digitisation of the energy system transformation plays a central role in this context. With the help of intelligent measuring systems, such as the smart meter and the SMGW required for communication, data on power generation and consumption can be recorded, transmitted and processed in real time (Maier 2018, p. 3). In accordance with § 29 Para. 3 MsbG, the Smart Meters are to be installed gradually and nationwide for all consumers by the year 2032. Larger consumers and generation plants are taking on a pioneering role in the use of modern measurement and control technology. Smaller electricity consumers will follow later. The SMGW is regarded as the key technology for digitizing the energy revolution. It communicates with various components and participating market players (BSI, no year) for the transmission of consumption data as well as for its administration. The SMGW guarantees data protection and data security at the highest level and lays the foundation for an intelligent and secure network. With the first certified SMGW, the BMWi and the BSI show that digitalization can succeed even with high data protection and information security requirements (Windmesse, 2018). In addition to modern measuring equipment and the control of generation and consumption facilities, blockchain technology also offers promising solutions and could prove revolutionary due to its properties (Zimmermann and Hoppe 2018, p. 13). The core functionality of the blockchain is the decentralized storage and encryption of transaction data in a long chain of data blocks. Basically, different types of blockchains can be used, which can be classified in terms of data access and network usage. In addition, there are different consensus mechanisms, the use of which depends strongly on the respective use case and the trust in the blockchain network. Blockchain technology has been further developed in recent years, resulting in new application possibilities. In particular, Smart-Contracts, which contain predefined transaction game rules with which contracts are automatically concluded and business processes can thus be made more efficient and simpler, contribute to this. The most important arguments for the use of blockchain technology in the energy market include simplified and automated processes, greater transparency and the reduction of transaction costs through disintermediation. However, there are also some arguments against the use of blockchain technology in the energy market. These include in particular the low transaction speed, illegal activities as well as the consumption of energy and resources. In addition, there are fundamental legal issues such as data protection. According to many estimates by experts from the energy sector, blockchain technology has the potential to significantly influence the energy industry in the coming years and thus unfold a new dynamic for the energy transition. A survey conducted by dena among 70 decision-makers in the energy sector in 2016 shows that the majority (69%) of respondents heard of blockchain applications in the energy sector. More than a third (39%) are already planning blockchain applications and 13% are already doing so. In addition, 60% agree that blockchain is likely to become more widespread in the energy sector. 21% of the respondents even see the blockchain as a "game changer" for the energy industry. A recent survey conducted this year by dena among 300 managers and experts in the energy industry in Germany, Austria and Switzerland shows that the majority (37%) of respondents are aware of the Blockchain's existence. 10% of the respondents had already implemented applications based on the blockchain. The applications are very diverse. Many of the use cases mentioned are automated charging and billing in the field of electromobility or proof of origin or certification of the origin and authenticity of green electricity. By far the most discussed use case, however, can be traced back to P2P trading, which makes electricity trading between private individuals economically possible without a participating energy company and thus also for prosumers despite low performance.

For this reason, this particular use case was examined in more detail within the framework of a concept. A potential assessment shows that P2P electricity trading can be combined in different ways. The decentralised P2P trade, which enables a new form of interaction, is particularly promising. Initial projects such as the "Brooklyn Microgrid" confirm the technical suitability of the blockchain for this particular area. Thanks to the decentralised transaction and energy supply system, which allows direct interactions between peers, intermediaries such as electricity suppliers and electricity traders/exchanges could be eliminated. An analysis of the regulatory framework shows, however, that such a concept for prosumers is only possible under difficult energy industry and bureaucratic conditions and can therefore be classified as unacceptable. This could be remedied by a service model in which all regulatory obligations are assumed by the platform user as a service provider. This could create a legally compliant electricity trade between prosumers and consumers, which would nevertheless make use of the advantages of blockchain technology. A business model, which was developed independently, comprises an infrastructure intended for this purpose, a detailed description of the individual process steps within the framework of a business process map as well as a potential design of the blockchain. In order to enable secure electricity trading between prosumers and consumers, a trading platform is used which is operated by a service provider (e.g. electricity supplier). From a technical point of view, some assumptions have been made. In addition to communication-capable hardware (Smart-Meter/SMGW), which is compatible with the blockchain, the possession of an electricity storage device is also a prerequisite. In order to map physical quantities such as the storage capacity (in kWh) on the blockchain, an EC token is assigned for each kWh, which also serves as proof of origin. Payment is made on the trading platform with EH tokens, which have a fixed equivalent value for reasons of value stability and are purchased by consumers using Fiat money. Prosumers who sell their generated electricity via the trading platform receive the agreed number of EH tokens credited. In principle, it is possible to purchase electricity via the platform yourself or to have the corresponding equivalent value paid out in Fiat money. The entire payment process can be automated and carried out securely using Smart-Contracts. Smart-Contracts also offer convincing added value for the checking of credit balances or the exclusion of a double sale. Due to the energy-efficient validation of transactions, data protection compliance, limited access and the resulting high transaction speeds, a consortium blockchain using the POA consensus mechanism is preferred. In order to check whether the business model also makes economic sense, a cost-benefit analysis is recommended. If the result of the analysis is an economic added value, the next step is to implement the business model as part of a proof-of-concept.

The blockchain is highly likely to change processes and business models in the energy sector in the long term. It remains to be seen how long and to what extent this will happen. The blockchain has the potential to accelerate energy transformation, rationalize processes and guarantee a higher level of data security and validity. However, the blockchain cannot replace a sustainable energy policy. The current regulatory framework is not designed for applications based on a blockchain. It must therefore be resolved how blockchain technology can be integrated into the existing regulatory framework. It must also be clarified how smart contracts can be combined with legally valid contracts. The Federal Government should create framework conditions which make use of the opportunities presented by the blockchain and address the challenges in an appropriate manner. With increasing public and commercial interest as well as the development of further projects, it can be assumed that the legal and technical challenges will be solved in the future on the basis of increasing experience. In future, it is to be expected that user-friendliness will be optimised, transaction speeds and costs reduced, data protection improved, interoperability created and energy consumption drastically reduced by alternative consensus mechanisms. From today's perspective, concrete applications are still difficult to implement due to the challenges mentioned above. Against this background, it can be assumed that - as promising as this technology may seem - no significant progress with regard to implemented applications can be expected in the foreseeable future. In principle, the earlier the regulatory and technical requirements for the use of the blockchain in the energy sector are met, the more consistently these added values can be tested and increased. With regard to P2P trading, it can be assumed that energy companies will increasingly offer services in the future as soon as they have recognised the added value. Should the legal framework develop in the right direction, it is conceivable that even the operators of a trading platform will become superfluous, so that producers and consumers can trade electricity autonomously among themselves. It is therefore to be hoped that in future it will be possible to overcome the regulations and the technical challenges that still exist.

Literature

1-Stromvergleich, 2018a. *Strompreisentwicklung in Deutschland: Die Strompreise 2019.* [Online]
Available at: https://1-stromvergleich.com/strompreisentwicklung/#2019
[accessed 26 November 2018].

1-Stromvergleich, 2018b. *Netzentgelte Strom: Hintergrund, Stand und Entwicklung bis 2019.* [Online]
Available at: https://1-stromvergleich.com/strom-report/netzentgelte/
[accessed 26 November 2018].

Agentur für Erneuerbare Energien, 2014. *Energiewende im Verkehrssektor kommt nur langsam in Gang.* [Online]
Available at: https://www.unendlich-viel-energie.de/energiewende-im-verkehrssektor-kommt-nur-langsam-in-gang
[accessed 29 October 2018].

Agentur für Erneuerbare Energien, 2018a. *DIE ENERGIEWENDE AUF DIE STRASSE BRINGEN.* [Online]
Available at: https://www.unendlich-viel-energie.de/media/file/1811.AEE_Renews_84_EW_auf_Strasse_bringen_Jan18-web.pdf
[accessed 1 December 2018].

Agentur für Erneuerbare Energien, 2018b. *Bürgerenergie bleibt Schlüssel für erfolgreiche Energiewende.* [Online]
Available at: https://www.unendlich-viel-energie.de/buergerenergie-bleibt-schluessel-fuer-erfolgreiche-energiewende
[accessed 2 December 2018].

Agentur für Erneuerbare Energien, no year. *Sektorenkopplung.* [Online]
Available at: https://www.unendlich-viel-energie.de/themen/strom/sektorenkopplung
[accessed 6 October 2018].

Agora Energiewende, 2017. *Energiewende und Dezentralität. Zu den Grundlagen einer politisierten Debatte..* [Online]
Available at: https://www.agora-energiewende.de/fileadmin2/Projekte/2016/Dezentralitaet/Agora_Dezentralitaet_WEB.pdf
[accessed 29 December 2018].

Aichele, C., 2011. *Smart Energy.* Ketsch: Springer.

Aichele, C. & Doleski, O., 2014. *Smart Market.* Wiesbaden: Springer.

Alexandre, A., 2018. *Skalierbarkeit: Blockchain kann Tagesvolumen des Aktienmarktes verarbeiten.* [Online]
Available at: https://de.cointelegraph.com/news/scalability-study-dlt-can-support-daily-trading-volume-of-us-equity-market
[accessed 11 January 2019].

Andersen, N., 2016. *Vorstellung der Blockchain-Technologie.* [Online]
Available at:
https://www2.deloitte.com/content/dam/Deloitte/de/Documents/Innovation/Vorstellung%20der%20Blockchain-Technologie.pdf
[accessed 3 January 2019].

Andoni, M. et al., 2019. *Blockchain technology in the energy sector: A systematic review of challenges and opportunities.* [Online]
Available at:
https://reader.elsevier.com/reader/sd/pii/S1364032118307184?token=48D89D4C1F252A110DA66081DF01FCBFC716FFA26BD51E88E881F46C1A1D2660FB85A7A046D3E80A2E9B83D886D7F020
[accessed 16 March 2019].

Arbeitsgemeinschaft für sparsamen und umweltfreundlichen Energieverbrauch, 2017. *Das KWK-Gesetz 2017.* [Online]
Available at:
https://www.asue.de/sites/default/files/asue/themen/blockheizkraftwerke/2017/broschueren/asue_kwk_gesetz2017_309860.pdf
[accessed 3 November 2018].

Aurbach, T., 2018. *Dezentrale Energieversorgung mit Photovoltaik.* [Online]
Available at: https://www.n-ergie-solar.de/photovoltaik-ratgeber/dezentrale-energieversorung-solaranlagen/
[accessed 15 November 2018].

Axa, 2019. *Fizzy.* [Online]
Available at: https://fizzy.axa/en-gb
[accessed 25 January 2019].

BaFin, 2017. *Blockchain-Technologie.* [Online]
Available at:
https://www.bafin.de/DE/Aufsicht/FinTech/Blockchain/blockchain_node.html
[accessed 22 December 2018].

Bashir, I., 2018. *Mastering Blockchain: Distributed ledger technology, decentralization, and smart-contracts explained.* 2 Hrsg. Birmingham: Packt.

Bauknecht, D., Vogel, M. & Funcke, S., 2015. *Energiewende – Zentral oder dezentral?.* [Online]
Available at: https://www.oeko.de/oekodoc/2368/2015-534-de.pdf
[accessed 23 December 2018].

BBWind, no year. *Funktionsweise einer Windenergieanlage und Komponenten.* [Online]
Available at: https://www.bbwind.de/windenergie_faq/technik/001.php
[accessed 6 December 2018].

Bech, M. & Garratt, R., 2017. *Kryptowährungen von Zentralbanken.* [Online]
Available at: https://www.bis.org/publ/qtrpdf/r_qt1709f_de.pdf
[accessed 21 January 2019].

Beckhaus, P., 2002. *Dissertation zum Thema Simulation und Anlagenmanagement für dezentrale Energieversorgungssysteme.* [Online]
Available at: https://www.zbt-duisburg.de/fileadmin/user_upload/01-aktuell/05-publikationen/02-langschriften/dissertation-peter-beckhaus.

Beegy, 2015. *Dezentrale Energieversorgung – die Vor- und Nachteile.* [Online]
Available at: https://www.beegy.com/dezentrale-energieversorgung-die-vor-und-nachteile/
[accessed 3 December 2018].

Bergmann, C., 2017. *Warum die deutsche Energiewirtschaft auf Blockchain steht.* [Online]
Available at: https://bitcoinblog.de/2017/01/13/warum-die-deutsche-energiewirtschaft-auf-blockchain-steht/
[accessed 5 February 2019].

Berlin-Klimaschutz, 2016. *EcoPool – das virtuelle Kraftwerk der GASAG.* [Online]
Available at: https://www.berlin-klimaschutz.de/de/projekte/ecopool-das-virtuelle-kraftwerk-der-gasag
[accessed 22 November 2018].

Bhamrah, D., 2018. *Blockchain: Internet of Transaction.* New Delhi: BPB Publications.

Binance, 2018. *Was ist ein Blockchain Konsensusalgorithmus.* [Online]
Available at: https://www.binance.vision/de/blockchain/what-is-a-blockchain-consensus-algorithm
[accessed December 2018].

Blocher, W., 2018. *Stellungnahme zur öffentlichen Anhörung zum Thema „Blockchain" des Ausschusses Digitale Agenda.* [Online]
Available at:
https://www.bundestag.de/blob/580710/5f685f73d40cc6445fee692de4c0a04d/a-drs--19-23-025-blocher-data.pdf
[accessed 30 December 2018].

Blockchain-Basics, 2017. *Hashing.* [Online]
Available at: www.blockchain-basics.com/Hashing.html
[accessed 18 December 2018].

Blocklab, 2017. *Skalieren Blockchains? Sorgen und Lösungsansätze.* [Online]
Available at: https://site.blocklab.de/2017/Skalierung/
[accessed 27 January 2019].

BMWi, 2015a. *Die Digitalisierung der Energiewende.* [Online]
Available at: https://www.bmwi.de/Redaktion/DE/Artikel/Energie/digitalisierung-der-energiewende.html
[accessed 28 December 2018].

BMWi, 2015b. *Rolloutplan nach dem Entwurf des Messstellenbetriebsgesetzes.* [Online]
Available at: https://www.bmwi.de/Redaktion/DE/Infografiken/Alt/intelligente-netze-zaehler-rollout-uebersicht.html
[accessed 10 October 2018].

BMWi, 2016. *Was ist ein "Prosumer"?.* [Online]
Available at: https://www.bmwi-energiewende.de/EWD/Redaktion/Newsletter/2016/06/Meldung/direkt-erklaert.html
[accessed 28 December 2018].

BMWi, 2018a. *Erneuerbare Energien.* [Online]
Available at: https://www.bmwi.de/Redaktion/DE/Dossier/erneuerbare-energien.html
[accessed 28 November 2018].

BMWi, 2018b. *Gesetzeskarte für das Energieversorgungssystem.* [Online]
Available at:
https://www.bmwi.de/Redaktion/DE/Publikationen/Energie/gesetzeskarte.pdf?__blob=publicationFile&v=37
[accessed 16 October 2018].

BMWi, 2019. *Barometer Digitalisierung der Energiewende. Ein neues Denken und Handeln für die Digitalisierung der Energiewende Berichtsjahr 2018.* [Online]
Available at: https://www.bmwi.de/Redaktion/DE/Publikationen/Studien/barometer-digitalisierung-der-energiewende.pdf?__blob=publicationFile&v=20
[accessed 12 February 2019].

BMWi, no year. *Zweite Verordnung zur Änderung der Anreizregulierungsverordnung.* [Online]
Available at: https://www.bmwi.de/Redaktion/DE/Downloads/XYZ/zweite-verordnung-aenderung-anreizregulierung-bundesregierungsverordnung.pdf?__blob=publicationFile&v=4
[accessed 17 December 2018].

Bogensperger, A., Zeiselmair, A. & Hinterstocker, M., 2018a. *DIE BLOCKCHAIN-TECHNOLOGIE CHANCE ZUR TRANSFORMATION DER ENERGIEVERSORGUNG? (TEIL: TECHNOLOGIEBESCHREIBUNG).* [Online]
Available at:
https://www.ffe.de/attachments/article/803/Blockchain_Teilbericht_Technologiebeschreibung.pdf
[accessed 20 January 2019].

Bogensperger, A., Zeiselmair, A., Hinterstocker, M. & Dufter, C., 2018b. *DIE BLOCKCHAIN-TECHNOLOGIE CHANCE ZUR TRANSFORMATION DER ENERGIEWIRTSCHAFT? (TEIL: ANWENDUNGSFÄLLE).* [Online]
Available at:
https://www.ffe.de/attachments/article/846/Blockchain_Teilbericht_UseCases.pdf
[accessed 13 February 2019].

Bollinger-Kanne, J., 2017. *Digitaler Energiewendeschub.* [Online]
Available at: https://www.vdi-nachrichten.com/Technik/Digitaler-Energiewendeschub
[accessed 16 March 2019].

Brauner, G., 2016. *Energiesysteme: regenerativ und dezentral.* Wien: Springer.

Breitsprecher, M., 2018. *Welches Problem löst die Blockchain?.* [Online]
Available at: https://blockruption.com/2018/06/welches-problem-lost-die-blockchain/
[accessed 15 December 2018].

BSI, 2016. *Bekanntmachung zur elektronischen Signatur nach dem Signaturgesetz und der Signaturverordnung.* [Online]
Available at:
https://www.bsi.bund.de/SharedDocs/Downloads/DE/BSI/ElekSignatur/Algorithmenkatalog2017_Entwurf.pdf?__blob=publicationFile&v=4
[accessed 22 December 2018].

BSI, 2018. *Das Smart-Meter-Gateway.* [Online]
Available at:
https://www.bsi.bund.de/SharedDocs/Downloads/DE/BSI/Publikationen/Broschueren/Smart-Meter-Gateway.pdf?__blob=publicationFile&v=6
[accessed 9 March 2019].

BSI, no year. *Smart Meter Gateway.* [Online]
Available at:
https://www.bsi.bund.de/DE/Themen/DigitaleGesellschaft/SmartMeter/SmartMeterGateway/smartmetergateway_node.html
accessed 29 December 2018].

Bundesamt für Wirtschaft und Ausfuhrkontrolle, 2018. *Kraft-Wärme-Kopplung.* [Online]
Available at:
http://www.bafa.de/DE/Energie/Energieeffizienz/Kraft_Waerme_Kopplung/Stromverguetung/stromverguetung_node.html
[accessed 5 November 2018].

Bundesministerium für Bildung und Forschung, no year. *Energiewende und nachhaltiges Wirtschaften.* [Online]
Available at: https://www.bmbf.de/de/energiewende-565.html
[accessed 3 October 2018].

Bundesministerium für Wirtschaft und Energie, 2015. *Die Energie der Zukunft. Vierter Monitoring-Bericht zur Energiewende..* [Online]
Available at: https://www.bmwi.de/Redaktion/DE/Publikationen/Energie/vierter-monitoring-bericht-energie-der-zukunft.pdf?__blob=publicationFile&v=24
[accessed 22 October 2018].

Bundesministerium für Wirtschaft und Energie, 2016. *Fit für den Strommarkt. Fit für die Zukunft. Alle wichtigen Fakten zum neuen EEG 2017.* [Online]
Available at: https://www.erneuerbare-energien.de/EE/Redaktion/DE/Downloads/Broschuere/fit-fuer-den-strommarkt-eeg-2017.pdf?__blob=publicationFile&v=5
[accessed 10 October 2018].

Bundesministerium für Wirtschaft und Energie, no year. *Solarenergie.* [Online]
Available at: https://www.erneuerbare-energien.de/EE/Navigation/DE/Technologien/Solarenergie-Photovoltaik/solarenergie-photovoltaik.html
[accessed 25 November 2018].

Bundesnetzagentur, 2011. *„Smart Grid" und „Smart Market" Eckpunktepapier der Bundesnetzagentur zu den Aspekten des sich verändernden Energieversorgungssystems.* [Online]
Available at:
https://www.bundesnetzagentur.de/SharedDocs/Downloads/DE/Sachgebiete/Energie/Unternehmen_Institutionen/NetzzugangUndMesswesen/SmartGridEckpunktepapier/SmartGridPapierpdf.pdf?__blob=publicationFile&v=2
[accessed 29 December 2018].

Bundesnetzagentur, no year. *Netzentgelt.* [Online]
Available at:
www.bundesnetzagentur.de/SharedDocs/FAQs/DE/Sachgebiete/Energie/Verbraucher/

Energielexikon/Netzentgelt.html?nn=267092rompreiszusammensetzung/l
[accessed 15 December 2018].

Bundesregierung, 2018a. *Koalitionsvertrag zwischen CDU, CSU und SPD 19. Legislaturperiode.* [Online]
Available at: https://www.bundesregierung.de/resource/blob/975226/847984/5b8bc23590d4cb2892b31c987ad672b7/2018-03-14-koalitionsvertrag-data.pdf?download=1
[accessed 5 February 2019].

Bundesregierung, 2018b. *Antwort der Bundesregierung auf die Kleine Anfrage der Abgeordneten Ingrid Nestle, Tabea Rößner, Dieter Janecek, weiterer Abgeordneter und der Fraktion BÜNDNIS 90/DIE GRÜNEN – Drucksache 19/4823 –.* [Online]
Available at: http://dip21.bundestag.de/dip21/btd/19/056/1905641.pdf
[accessed 5 February 2019].

Bundesverband der Energie- und Wasserwirtschaft, 2017. *Blockchain in der Energiewirtschaft.* [Online]
Available at: https://www.bdew.de/media/documents/BDEW_Blockchain_Energiewirtschaft_10_2017.pdf
[accessed 20 January 2019].

Bundesverband der Energie- und Wasserwirtschaft, 2018a. *BDEW-Strompreisanalyse Mai 2018.* [Online]
Available at: https://www.bdew.de/media/documents/1805018_BDEW-Strompreisanalyse-Mai-2018.pdf
[accessed 22 November 2018].

Bundesverband der Energie- und Wasserwirtschaft, 2018b. *Entwicklung der Strompreise.* [Online]
Available at: https://www.bdew.de/presse/pressemappen/entwicklung-der-strompreise/
[accessed 25 November 2018].

Bundesverband Erneuerbare Energie, 2018. *Klare Zielverfehlung bei EU-Verpflichtung zum Ausbau Erneuerbarer Energien in Deutschland.* [Online]
Available at: https://www.bee-ev.de/home/presse/mitteilungen/detailansicht/klare-zielverfehlung-bei-eu-verpflichtung-zum-ausbau-erneuerbarer-energien-in-deutschland/
[accessed 30 November 2018].

Bundesverband Informationswirtschaft, Telekommunikation und neue Medien, 2017. *Blockchain und Datenschutz.* [Online]
Available at: https://www.bitkom.org/sites/default/files/file/import/180502-Faktenpapier-

Blockchain-und-Datenschutz.pdf
[accessed 15 January 2019].

Bunnemann, M., 2018. *Digitalisierung als Treiber der urbanen Energiewende.* [Online]
Available at: https://www.eon.com/de/neue-energie/digitalisierung-treiber-urbaner-energiewende.html
[accessed 4 February 2019].

Burger, C., Trbovich, A. & Weinmann, J., 2018. *Vulnerabilities in smart meter infrastructure – can blockchain provide a solution?.* [Online]
Available at:
http://shop.dena.de/fileadmin/denashop/media/Downloads_Dateien/esd/9236_Vulnerabilities_in_smart_meter_infrastructure_-can_blockchain_provide_a_solution.pdf
[accessed 5 February 2019].

Burgwinkel, D., 2016. *Blockchain Technology.* Basel: DE GRUYTER OLDENBOURG.

Coinmarketcap, 2019a. *Nxt.* [Online]
Available at: https://coinmarketcap.com/currencies/nxt/
[accessed 15 January 2019].

Coinmarketcap, 2019b. *Coinmarketcap.* [Online]
Available at: https://www.coinmarketcap.com
[accessed 18 January 2019].

Cointelegraph, no year. *Was ist Ethereum? Ein Leitfaden für Anfänger.* [Online]
Available at: https://de.cointelegraph.com/ethereum-for-beginners/what-is-ethereum
[accessed 2 January 2019].

Condos, J., Sorrell, W. & Donegan, S., 2016. *BLOCKCHAIN TECHNOLOGY: OPPORTUNITIES AND RISKS.* [Online]
Available at: https://legislature.vermont.gov/assets/Legislative-Reports/blockchain-technology-report-final.pdf
[accessed 5 December 2018].

CPMI, 2015. *Digital currencies.* [Online]
Available at: https://www.bis.org/cpmi/publ/d137.pdf
[accessed 5 January 2019].

Cryptolist, no year. *Was ist eine Blockchain?.* [Online]
Available at: https://www.cryptolist.de/was-ist-blockchain
[accessed 10 January 2019].

Dannen, C., 2017. *Introducing Ethereum and Solidity: Foundations of Cryptocurrency and Blockchain Programming for Beginners.* New York: Apress.

dena, 2016. *Blockchain: Energiewirtschaft bereitet sich auf neues digitales Verfahren für Transaktionen vor.* [Online]

Available at: https://www.dena.de/newsroom/meldungen/blockchain-energiewirtschaft-bereitet-sich-auf-neues-digitales-verfahren-fuer-transaktionen-vor/
[accessed 5 February 2019].

dena, 2019a. *Energiewirtschaft erprobt Blockchain in der Praxis.* [Online]
Available at: https://www.dena.de/newsroom/meldungen/energiewirtschaft-erprobt-blockchain-in-der-praxis/
[accessed 5 February 2019].

dena, 2019b. *Blockchain in der integrierten Energiewende.* [Online]
Available at: https://www.dena.de/newsroom/publikationsdetailansicht/pub/blockchain-in-der-integrierten-energiewende/
[accessed 4 March 2019].

Denis, T. & Johnson, S., 2017. *Kryptografie für Entwickler: Das erste umfassende Kryptografie-Handbuch für Software-Entwickler.* Ottawa: Franzis.

Deutsche Energie-Agentur, 2016. *Potenzielle Anwendungsfelder der Blockchain in der Energiewende.* [Online]
Available at:
https://www.dena.de/fileadmin/dena/Bilder/Newsroom/Meldungen/2016/grafik_blockchain_de_print.jpg
[accessed 5 February 2019].

Deutsche Energie-Agentur, 2019. *Umfrage zur Blockchain in der Energiewirtschaft.*
[Online]
Available at:
https://www.dena.de/fileadmin/dena/Dokumente/Presse___Medien/Hintergrundpapier_Blockchain_Umfrage_final.pdf
[accessed 10 February 2019].

Deutsche Energie-Agentur, no year. *Energiespeicher zur flexiblen Integration erneuerbarer Energien.* [Online]
Available at: https://www.dena.de/themen-projekte/energiesysteme/flexibilitaet-und-speicher/batteriespeicher/
[accessed 10 December 2018].

Deutscher Bundestag, 2018. *Fragen zu Blockchain und Kryptowährungen.* [Online]
Available at:
https://www.bundestag.de/blob/554992/05906ed0058d179d1eed6086be937bfb/wd-4-051-18-pdf-data.pdf
[accessed 27 January 2019].

Deutscher Bundestag, 2019. *Die Rolle der Blockchain-Technologie in der Energiewirtschaft.* [Online]

Available at: http://dip21.bundestag.de/dip21/btd/19/072/1907286.pdf
[accessed 30 January 2019].

Doelling, R., 2016. *Kann die Blockchain der Energiewende und dem Strommarkt helfen?.* [Online]
Available at: https://blog.energiedienst.de/blockchain-strommarkt/
[accessed 4 February 2019].

Doleski, O., 2017. *Herausforderung Utility 4.0: Wie sich die Energiewirtschaft im Zeitalter der Digitalisierung verändert.* Ottobrunn: Springer.

Drømstørre, 2003. *Getriebe für Windkraft- anlagen.* [Online]
Available at: http://drømstørre.dk/wp-content/wind/miller/windpower%20web/de/tour/wtrb/powtrain.htm
[accessed 5 December 2018].

Drescher, D., 2017. *Blockchain Grundlagen. Eine Einführung in die elementaren Konzepte in 25 Schritten.* Frechen: Mitp.

Duden, no year. *Konsens.* [Online]
Available at: https://www.duden.de/rechtschreibung/Konsens
[accessed 27 December 2018].

Eckstein, A., Liebetrau, A. & Funk-Münchmeyer, A., 2018. *Insurance & Innovation 2018: Ideen und Erfolgskonzepte von Experten aus der Praxis.* Karlsruhe: VVW.

Egloff, P. & Turnes, E., 2019. *Blockchain für die Praxis.* Zürich: SKV.

EHA, 2018. *Der Strommarkt in Deutschland – Überblick & Akteure des Strommarktes.* [Online]
Available at: https://www.eha.net/blog/details/strommarkt-deutschland.html
[accessed 13 March 2019].

Ehrlich, A., 2017. *Enerchain - Energiegroßhandel auf der Blockchain.* [Online]
Available at: https://www.energie-und-management.de/nachrichten/detail/enerchain-energiegrosshandel-auf-der-blockchain-121167
[accessed 4 February 2019].

Elect-Expo, no year. *Blockchain: Lösung für den dezentralen Mobilitätsmarkt.* [Online]
Available at: https://www.elect-expo.com/electrifying-insights/blockchain-loesung-fuer-den-dezentralen-mobilitaetsmarkt/
[accessed 27 February 2019].

Energie-Experten, 2016. *Technik und Einsatz eines Stromspeichers in Solaranlagen.* [Online]
Available at: https://www.energie-experten.org/erneuerbare-energien/photovoltaik/stromspeicher.html
[accessed 17 November 2018].

Energie-Forschungszentrum Niedersachsen, 2013. *Studie: Eignung von Speichertechnologien zum Erhalt der Systemsicherheit.* [Online]
Available at: https://www.bmwi.de/Redaktion/DE/Publikationen/Studien/eignung-von-speichertechnologien-zum-erhalt-der-systemsicherheit.pdf?__blob=publicationFile&v=10
[accessed 15 November 2018].

Energie-Lexikon, 2018a. *Dezentrale Energieerzeugung.* [Online]
Available at: https://www.energie-lexikon.info/dezentrale_energieerzeugung.html
[accessed 2 December 2018].

Energie-Lexikon, 2018b. *Sonnenenergie.* [Online]
Available at: https://www.energie-lexikon.info/sonnenenergie.html
[accessed 2 December 2018].

Energie-Lexikon, 2018c. *Photovoltaik.* [Online]
Available at: https://www.energie-lexikon.info/photovoltaik.html
[accessed 2 December 2018].

Energie-Lexikon, 2018d. *Windenergieanlage.* [Online]
Available at: https://www.energie-lexikon.info/windenergieanlage.html
[accessed 2 December 2018].

Energyweb, no year. *Energyweb.* [Online]
Available at: https://energyweb.org
[accessed 27 December 2018].

Eon, no year. *Digitalisierung der Energiewende: Alles hängt zusammen.* [Online]
Available at: https://www.eon.de/de/eonerleben/digitalisierung-der-energiewende-gesetz.html
[accessed 28 December 2018].

Erneuerbar Mobil, 2016. *Wozu gibt es das Elektromobilitätsgesetz?.* [Online]
Available at: https://www.erneuerbar-mobil.de/faq/wozu-gibt-es-das-elektromobilitaetsgesetz
[accessed 27 October 2018].

Ethereum Homestead, 2016. *Developer Tools.* [Online]
Available at: http://ethdocs.org/en/latest/contracts-and-transactions/developer-tools.html?highlight=language
[accessed 24 January 2019].

Ethereum-base, 2017. *Ethereum-base.* [Online]
Available at: http://ethereum-base.com/blockchain/
[accessed 2 January 2019].

Etherworld, 2017. *Concept of the Merkle Tree in Ethereum.* [Online]
Available at: https://etherworld.co/2017/02/14/concept-of-merkle-tree-in-ethereum/

Fören, no year. *Batteriespeicher.* [Online]
Available at: https://www.foeren.de/erneuerbare/energiespeicher/batteriespeicher/
[accessed 16 November 2018].

FfE, 2018. *Neue Studie: Anwendungsfälle der Blockchain-Technologie in der Energiewirtschaft.* [Online]
Available at: https://www.ffe.de/themen-und-methoden/digitalisierung/846-chancen-der-blockchain-technologie-in-der-energiewirtschaft-anwendungsfaelle.html
[accessed 10 February 2019].

Fischer, C., Fiedler, I. & Babenko, L., 2019. Blockchain-Technologie im Handel der Zukunft. In: G. Heinemann, M. Gehrckens & T. Täuber, Hrsg. *Handel mit Mehrwert: Digitaler Wandel in Märkten, Geschäftsmodellen und Geschäftssystemen.* Wiesbaden: Springer.

Focus, 2018. *Steve Wozniak: „Nur Bitcoin ist pures digitales Gold".* [Online]
Available at: https://www.focus.de/finanzen/boerse/kryptowaehrungen/apple-mitgruender-steve-wozniak-steve-wozniak-nur-bitcoin-ist-pures-digitales-gold_id_9046928.html
[accessed 21 January 2019].

ForschungsVerbund Erneuerbare Energien, 2010. *Energiekonzept 2050.* [Online]
Available at:
http://www.fvee.de/fileadmin/politik/10.06.vision_fuer_nachhaltiges_energiekonzept.pdf
[accessed 01 November 2018].

Fraunhofer, 2019. *Aktuelle Fakten zur Photovoltaik in Deutschland.* [Online]
Available at:
https://www.ise.fraunhofer.de/content/dam/ise/de/documents/publications/studies/aktuelle-fakten-zur-photovoltaik-in-deutschland.pdf
[accessed 18 March 2019].

Frondel, M., Schmidt, C. & aus dem Moore, N., 2012. *Cesifo Group.* [Online]
Available at: https://www.cesifo-group.de/DocDL/ifosd_2012_17_1.pdf
[accessed 12 November 2018].

Günther, M., 2015. *Energieeffizienz durch Erneuerbare Energien. Möglichkeiten, Potenziale, Systeme.* Kassel: Springer.

Günther, S., 2016. *Strom selbst erzeugen und selbst verbrauchen – Informationen zur Photovoltaik-Lösung.* [Online]
Available at: https://www.energieheld.de/blog/strom-erzeugen-und-verbrauchen/
[accessed 20 November 2018].

Geiselhardt, S., 2017. *Kann Blockchain den Energiesektor revolutionieren?*. [Online]
Available at: http://www.sonnewindwaerme.de/panorama/blockchain-energiesektor-revolutionieren
[accessed 11 February 2019].

Glaser, F. & Bezzenberger, L., 2015. *BEYOND CRYPTOCURRENCIES: A TAXONOMY OF DECENTRALIZED CONSENSUS SYSTEMS.* [Online]
Available at: https://balsa.man.poznan.pl/indico/event/44/material/paper/0?contribId=237
[accessed 7 December 2018].

Glatz, F., 2018. *Stellungnahme des Blockchain Bundesverband.* [Online]
Available at: https://www.bundestag.de/blob/580950/6f592a83b376199a092e1616eaba5402/a-drs--19-23-028-glatz-data.pdf
[accessed 27 January 2019].

Gochermann, J., 2016. *Expedition Energiewende.* Wiesbaden: Springer.

Goldin, M., 2016. *Medium.* [Online]
Available at: https://medium.com/@ConsenSys/ethereum-bitcoin-plus-everything-a506dc780106
[accessed 23 January 2019].

Gründiger, W., no year. *Smart Home.* [Online]
Available at: https://www.bvdw.org/der-bvdw/gremien/smart-home/news/
[accessed 12 November 2018].

Graulich, K. et al., 2018. *Einsatz und Wirtschaftlichkeit von Photovoltaik-Batteriespeichern in Kombination mit Stromsparen.* [Online]
Available at: https://www.oeko.de/fileadmin/oekodoc/PV-Batteriespeicher-Endbericht.pdf
[accessed 20 November 2018].

Gridx, 2017. *Dezentrale Energieversorgung – Die Idee hinter gridX.* [Online]
Available at: https://gridx.de/2017/12/03/dezentrale-energieversorgung/
[accessed 26 December 2018].

Hannen, P., 2018. *Unternehmen treiben Blockchain im Energiehandel voran.* [Online]
Available at: https://www.pv-magazine.de/2018/08/24/unternehmen-treiben-blockchain-im-energiehandel-voran/
[accessed 5 February 2019].

Hasse, F. et al., 2016. *Blockchain - Chance für die Energieverbraucher?.* [Online]
Available at: https://www.pwc.de/de/energiewirtschaft/blockchain-chance-fuer-

energieverbraucher.pdf
[accessed 02 January 2019].

Hein, C., Wellbrock, W. & Hein, C., 2019. *Rechtliche Herausforderungen von Blockchain-Anwendungen: Straf-, Datenschutz- und Zivilrecht.* Wiesbaden: SpringerGabler.

Heinze, Y. & Protschka, F., 2018. *Blockchain ABC: von A wie Altcoin bis Z wie ZCash.* Flensburg: epubli.

Heizung-Fachberater, 2018. *Welche Vor- und Nachteile hat eine dezentrale Energieversorgung?.* [Online]
Available at: http://heizung-fachberater.de/dezentrale-energieversorgung/
[accessed 25 December 2018].

Hohenberger, T. & Mühlenhoff, J., 2014. *Agentur für Erneuerbare Energien (AEE).* [Online]
Available at: https://www.unendlich-viel-energie.de/media/file/320.71_Renews_Spezial_Energiewende_im_Verkehr_online_apr14.pdf
[accessed 18 October 2018].

Holstenkamp, L. & Radtke, J., 2018. *Handbuch Energiewende und Partizipation.* Wiesbaden: Springer.

Hosp, J., 2018. *Blockchain 2.0: einfach erklärt - weit mehr als nur Bitcoin.* s.l.:FinanzBuch Verlag.

Ingenieurbüro Welsch, no year. *Entwicklungsperspektiven dezentraler Energietechnologien im Strom- und Wärmemarkt.* [Online]
Available at: http://www.ib-welsch.de/Analyse_dezentrale_Energietechnologie.pdf
[accessed 2 December 2018].

ITWissen, 2018. *Smart Energy.* [Online]
Available at: https://www.itwissen.info/Smart-Energy-smart-energy.html
[accessed 28 December 2018].

IWR, 2018. *News.* [Online]
Available at: https://www.iwr.de/news.php?id=35533
[accessed 26 November 2018].

Jüttemann, P., 2017. *Unterschiede zwischen Solarstrom- und Kleinwindanlagen: Planungsfehler vermeiden.* [Online]
Available at: https://www.klein-windkraftanlagen.com/allgemein/unterschiede-zwischen-pv-anlagen-und-kleinwindenergieanlagen-kennen-planungsfehler-vermeiden/
[accessed 1 December 2018].

Jauch, D., 2017. *PV-Speichersysteme: Unterschied AC und DC.* [Online]
Available at: https://www.memodo.de/blog/pv-speichersysteme-unterschied-ac-und-dc/
[accessed 20 November 2018].

Kühl, A., 2017. *Was ist eigentlich ein Prosumer?*. [Online]
Available at: https://www.energynet.de/2017/04/10/prosumer/
[accessed January 2019].

Kühl, A., 2018. *Welche Rolle spielt die Digitalisierung in der Energiewende?*. [Online]
Available at: https://www.energynet.de/2018/09/05/digitalisierung-energiewende-2/
[accessed 26 November 2018].

Kühne, O. & Weber, F., 2018. *Bausteine der Energiewende*. Wiesbaden: Springer.

Kästner, T. & Kießling, A., 2016. *Energiewende in 60 Minuten. Reiseführer durch die Stromwirtschaft*. Wiesbaden: Springer.

Kalthofen, T. & Dose, J., 2018. *Die Grundlagen von Blockchain*. [Online]
Available at: https://www.channelpartner.de/a/die-grundlagen-von-blockchain,3330054
[accessed 27 January 2019].

Kfw, 2017. *Erneuerbare Energien – Standard*. [Online]
Available at: https://www.kfw.de/inlandsfoerderung/Unternehmen/Energie-Umwelt/Förderprodukte/Erneuerbare-Energien-Standard-(270)/
[accessed 25 November 2018].

Kleimaier, M. & Pokojski, M., 2008. *Dezentrale Energieversorgung - Ergebnisse der ETG-Studie*. [Online]
Available at:
http://lms.ee.hm.edu/~seck/AlleDateien/Allgemeines/VDE2008/Proceedings/ETG_2_1_2_Kleimaier.pdf
[accessed 12 December 2018].

Kloth, C., 2017a. *Bayerisches Wildpoldsried wird Blockchain-Labor*. [Online]
Available at: https://bizz-energy.com/blockchain_mekka_wildpoldsried
[accessed 4 February 2019].

Kloth, C., 2017b. *Wuppertaler Stadtwerke handeln Ökostrom über Blockchain*. [Online]
Available at: https://bizz-energy.com/wuppertaler_stadtwerke_handeln_oekostrom_ueber_blockchain
[accessed 4 February 2019].

Kloth, C., 2018a. *Verband: Smart Meter müssen Blockchain-fähig sein*. [Online]
Available at: https://bizz-energy.com/verband_smart_meter_muessen_blockchain_faehig_sein
[accessed 4 February 2019].

Kloth, C., 2018b. *Energiegroßhandel: Blockchain-Projekt Enerchain auf der Zielgeraden*.
[Online]
Available at: https://bizz-

energy.com/energiegrosshandel_blockchain_projekt_enerchain_auf_der_zielgeraden [accessed 16 March 2019].

Kloth, C., 2018c. *EWF senkt Stromverbrauch ihrer Energie-Blockchain.* [Online] Available at: https://bizz-energy.com/ewf_senkt_stromverbrauch_ihrer_energie_blockchain [accessed 16 March 2019].

Kuhlmann, A., Burger, C., Richard, P. & Weinmann, J., 2016. *Blockchain in der Energiewende. Eine Umfrage unter Führungskräften der deutschen Energiewirtschaft.* [Online] Available at: https://www.dena.de/fileadmin/dena/Dokumente/Pdf/9165_Blockchain_in_der_Energiewende_deutsch.pdf [accessed 5 February 2019].

Kulturwissenschaften, no year. *Projekt EnerDigit: Digitalisierung & Energiewende zwischen Dezentralität und Zentralität: regionale und unternehmenskulturelle Perspektiven.* [Online] Available at: http://www.kulturwissenschaften.de/cms/projekt-166.html [accessed 27 December 2018].

Lang, M. & Karlstetter, F., 2017. *Consensus-Modelle in der Übersicht.* [Online] Available at: https://www.dev-insider.de/consensus-modelle-in-der-uebersicht-a-631671/ [accessed 28 December 2018].

Laurence, T., 2017. *Blockchain für Dummies.* Weinheim: WILEY.

Lehrerfreund, 2011. *Stromerzeugung mit Windenergieanlagen.* [Online] Available at: https://www.lehrerfreund.de/technik/1s/stromerzeugung-mit-windenergieanlagen1/4042 [accessed 3 December 2018].

Lifestrom, 2017. *Welche Auswirkungen hat die Energiewende auf den Strompreis?.* [Online] Available at: https://www.lifestrom.de/magazin/welche-auswirkungen-hat-die-energiewende-auf-den-strompreis/ [accessed 26 November 2018].

Linden, C., 2018. *Immer mehr Anlagen fallen ab 2020 aus der EEG-Förderung.* [Online] Available at: https://www.pwc.de/de/energiewirtschaft/eeg-foerderung-fuer-alte-photovoltaik-anlagen-laeuft-aus.html [accessed 3 March 2019].

Lissek, S., 2016. *8. Göttinger Tagung: Braucht ein Verteilnetzbetreiber Systemdienstleistungen?.* [Online]

Available at:
https://www.efzn.de/fileadmin/documents/Goettinger_Energietagung/Vorträge/2016/04_Lissek.pdf
[accessed 3 December 2018].

Mücke, T., 2018. *Proof of Authority.* [Online]
Available at: https://www.coin-report.net/de/proof-of-authority/
[accessed 28 December 2018].

Müller, A., 2018. *Digitalisierung und Energiewende zusammendenken.* [Online]
Available at: https://www.unendlich-viel-energie.de/metaanalyse-digitalisierung-energiewende
[accessed 1 December 2018].

Maier, M., 2018. *Metaanalyse: Die Digitalisierung der Energiewende.* [Online]
Available at:
http://www.forschungsradar.de/fileadmin/content/bilder/Vergleichsgrafiken/meta_digitalisierung_aug18/AEE_Metanalyse_Digitalisierung_aug18.pdf
[accessed 28 November 2018].

Malanov, A., 2017. *Warum die Blockchain-Technologie gar nicht so schlecht ist.* [Online]
Available at: https://www.kaspersky.de/blog/good-good-blockchain/14870/
[accessed 11 January 2019].

Margaritoff, M., 2018. *The Scalability Trilemma.* [Online]
Available at: https://www.jeffersoncapital.info/the-scalability-trilemma/
[accessed 27 January 2019].

Mauro, A., 2006. *A pure P2P network.* [Online]
Available at: https://www.researchgate.net/figure/A-pure-P2P-network_fig1_34684930
[accessed 16 December 2018].

Meinel, C., Gayvoronskaya, T. & Schnjakin, M., 2017. *Blockchain: Hype oder Innovation.* Potsdam: Universitätsverlag Potsdam.

Meta-Level, no year. *Smart Energy – Mit intelligenten Ideen zur Energiewende.* [Online]
Available at: https://www.meta-level.de/smart-energy-mit-intelligenten-ideen-zur-energiewende/
[accessed 28 December 2018].

Mitschele, A., no year. *Blockchain.* [Online]
Available at: https://wirtschaftslexikon.gabler.de/definition/blockchain-54161
[accessed 5 December 2018].

Mullender, S., 1990. *Introduction to Distributed Systems.* [Online]
Available at: https://ris.utwente.nl/ws/files/6145134/mullender92introduction.pdf
[accessed 7 December 2018].

Nakamoto, S., 2008. *Bitcoin: A Peer-to-Peer Electronic Cash System.* [Online]
Available at: https://bitcoin.org/bitcoin.pdf
[accessed 3 January 2019].

Nees, F., 2018. *Blockchain: Die Vorteile und die Nachteile.* [Online]
Available at: https://www.cio.de/a/blockchain-die-vorteile-und-die-nachteile,3563566,6
[accessed 4 February 2019].

Neugebauer, R., 2018. *Digitalisierung: Schlüsseltechnologien für Wirtschaft & Gesellschaft.* München: Springer.

Pagel, U., 2018. *Entscheidungsbaum Blockchain: Anwendungsbeschreibung.* [Online]
Available at: https://blockchain-initiative.de/uncategorized/entscheidungsbaum-blockchain-anwendungsbeschreibung/
[accessed 15 February 2019].

Palka, S. & Wittpahl, V., 2018. *Vertrauen und Transparenz – Blockchain-Technologie als digitaler Vertrauenskatalysator.* [Online]
Available at: https://www.iit-berlin.de/de/publikationen/vertrauen-und-transparenz-blockchain-technologie-als-digitaler-vetrauenskatalysator/at_download/download
[accessed 27 January 2019].

Petri, T., 2018. *Kryptowährungen allgemein erklärt und die „Top 30" in der Übersicht.* Berlin: krypto-research@gmx.de.

Preiß, S., 2018a. *„Blockchain hat das Potenzial, die Energiewirtschaft in den kommenden Jahren maßgeblich zu beeinflussen".* [Online]
Available at: https://www.euwid-energie.de/blockchain-hat-das-potenzial-die-energiewirtschaft-in-den-kommenden-jahren-massgeblich-zu-beeinflussen/
[accessed 4 February 2019].

Preiß, S., 2018b. *Blockchain in der Energiewirtschaft und der Wunsch nach einem gesetzlichen Rahmen.* [Online]
Available at: https://www.euwid-energie.de/blockchain-in-der-energiewirtschaft-und-der-wunsch-nach-einem-gesetzlichen-rahmen/
[accessed 10 February 2019].

Preiß, S., 2018c. *Blockchain-Technologie bietet nicht per se für jeden Anwendungsfall Vorteile.* [Online]
Available at: https://www.euwid-energie.de/blockchain-technologie-bietet-nicht-per-se-fuer-jeden-anwendungsfall-vorteile
[accessed 7 March 2019].

Prognos, 2016. *Dezentralität und zellulare Optimierung – Auswirkungen auf den Netzausbaubedarf.* [Online]
Available at: https://www.n-

ergie.de/public/remotemedien/media/n_ergie/internet/die_n_ergie/unternehmen_1/dezentralitaetsstudie/N-ERGIE_Studie_Zellulare_Optimierung_final.pdf

Radtke, J. & Kersting, N., 2018. *Energiewende Politikwissenschaftliche Perspektiven.* Siegen: Springer.

Rauscher, A. & Cupic, Z., 2018. *Blockchain bassierte Smart Contracts: Grundlagen, Prozessunterstützung und Bewertung.* [Online]
Available at: https://www.syntax-solution.de/wp-content/uploads/sites/7/2018/02/Blockhain-bassierte-Smart-Contract-Andreas-Rauscher-Zoran-Cupic.pdf

Rechnerphotovoltaik, no year. *Die Funktionsweise einer Photovoltaikanlage.* [Online]
Available at: https://www.rechnerphotovoltaik.de/photovoltaik/technik/funktionsweise
[accessed 2 December 2018].

Reck, H.-J., 2011. *Kombikraftwerk: Stabiler Strom aus Erneuerbaren Energien.* [Online]
Available at: http://www.energiewende-sta.de/tag/erneuerbare-energien/page/9/
[accessed 24 December 2018].

Reiner Lemoine Institut, 2013. *Vergleich und Optimierung von zentral und dezentral orientierten Ausbaupfaden zu einer Stromversorgung aus erneuerbaren Energien in Deutschland.* [Online]
Available at: https://www.bvmw.de/fileadmin/pdf-archiv/Studie_zur_dezentralen_Energiewende.pdf.pdf
[accessed 1 December 2018].

Renn, O., no year. *Resilienz.* [Online]
Available at: https://energiesysteme-zukunft.de/themen/resilienz/
[accessed 1 December 2018].

Rentrop, C. & Augsten, S., 2017. *Was ist ein Smart Contract?.* [Online]
Available at: https://www.dev-insider.de/was-ist-ein-smart-contract-a-585679/
[accessed 23 January 2019].

Richard, P., Mamel, S. & Vogel, L., 2019. *Blockchain in der integrierten Energiewende.* [Online]
Available at: https://www.dena.de/fileadmin/dena/Publikationen/PDFs/2019/dena-Studie_Blockchain_Integrierte_Energiewende_DE4.pdf
[accessed 26 February 2019].

Rosenberger, P., 2018. *Bitcoin und Blockchain.* Münster: Springer.

Roth, I., 2018. *Digitalisierung in der Energiewirtschaft.* [Online]
Available at: https://www.boeckler.de/pdf/p_fofoe_WP_073_2018.pdf
[accessed 28 December 2018].

Sauer, D., 2015. *Ist die Zukunft der Energieversorgung zentral oder dezentral?.* [Online]
Available at:
https://www.researchgate.net/profile/Dirk_Sauer/publication/299388596_Ist_die_Zukunft_der_Energieversorgung_zentral_oder_dezentral_Ein_Diskussionsbeitrag_zu_Ursachen_Wirkungen_und_Ergebnissen_der_Entwicklung/links/56f2ff5208ae95e8b6cb4848/Ist-die-Zukunft-
accessed 20 November 2018].

Schütte, J., Fridgen, G., Prinz, W. & Rose, T., 2017. *BLOCKCHAIN UND SMART CONTRACTS.* [Online]
Available at:
https://www.sit.fraunhofer.de/fileadmin/dokumente/studien_und_technical_reports/Fraunhofer-Positionspapier_Blockchain-und-Smart-Contracts.pdf?_=1516641660
[accessed 3 November 2018].

Schütz, A. et al., 2018. *Vertrauen ist gut, Blockchain ist besser – Einsatzmöglichkeiten von Blockchain für Vertrauensprobleme im Crowdsourcing.* [Online]
Available at: https://doi.org/10.1365/s40702-018-00471-9
[accessed 23 January 2019].

Schabbach, T. & Wesselak, V., 2012. *Energie. Die Zukunft wird erneuerbar.* Nordhausen: Springer.

Schiller, K., 2018a. *Proof of Authority | Kompromiss zum Proof of Stake?.* [Online]
Available at: https://blockchainwelt.de/proof-of-authority-poa/
[accessed 25 December 2018].

Schiller, K., 2018b. *Was sind Smart Contracts? | Definition und Erklärung.* [Online]
Available at: https://blockchainwelt.de/smart-contracts-vertrag-blockchain/
[accessed 10 January 2019].

Schiller, K., 2018c. *Solidity | Programmiersprache von Ethereum.* [Online]
Available at: https://blockchainwelt.de/solidity-smart-contracts-programmieren/
[accessed 20 January 2019].

Schlatt, V., Schweizer, A., Urbach, N. & Fridgen, G., 2016. *Blockchain: Grundlagen, Anwendungen und Potenziale.* [Online]
Available at:
https://www.fit.fraunhofer.de/content/dam/fit/de/documents/Blockchain_WhitePaper_Grundlagen-Anwendungen-Potentiale.pdf
[accessed 21 January 2019].

Schreier, J., 2017. *Blockchain wird 2018 das IoT revolutionieren.* [Online]
Available at: https://www.industry-of-things.de/blockchain-wird-2018-das-iot-

revolutionieren-a-671563/index4.html
[accessed 5 February 2019].

Schwan, G., Treichel, K. & Höh, A., 2016. *HUMBOLDT-VIADRINA Governance Platform.*
[Online]
Available at: https://www.governance-platform.org/wp-content/uploads/2017/03/HVGP_Trialog-Bericht-Sektorkopplung.pdf
[accessed 12 October 2018].

Seidel, M., 2019. *Banking & Innovation 2018/2019.* Stuttgart: Springer Gabler.

Servatius, H.-G., Schneidewind, U. & Rohlfing, D., 2012. *Smart Energy.* Heidelberg: Springer.

Siegel, D. & Andersen, N., no year. *Blockchain – ein Game-Changer?.* [Online]
Available at: https://www2.deloitte.com/de/de/pages/innovation/contents/Blockchain-Game-Changer.html#
[accessed 5 February 2019].

Siepermann, M., no year. *Smart City.* [Online]
Available at: https://wirtschaftslexikon.gabler.de/definition/smart-city-54505
[accessed 25 November 2018].

Sieverding, U. & Schneidewindt, H., 2016. *BLOCKCHAIN IN DER ENERGIEWIRTSCHAFT.* [Online]
Available at: http://library.fes.de/pdf-files/wiso/12996.pdf
[accessed 6 February 2019].

Sinegal, J., 2018. *Blockchain: Möglichkeiten und Grenzen der dezentralisierten Technologie.*
[Online]
Available at: http://www.morningstar.de/de/news/168852/blockchain-möglichkeiten-und-grenzen-der-dezentralisierten-technologie.aspx
[accessed 15 January 2019].

Sixt, E., 2016. *Bitcoins und andere dezentrale Transaktionssysteme.* Wien: Springer.

Skvorc, B. et al., 2018. *Learn Ethereum: The Collection.* s.l.:Sitepoint.

Smartcityduisburg, 2017. *Smart City.* [Online]
Available at: https://www.duisburgsmartcity.de
[accessed 28 November 2018].

Solaranlage.eu, no year. *Funktionsweise der Photovoltaikanlage.* [Online]
Available at: https://www.solaranlage.eu/photovoltaik/technik-komponenten/funktionsweise-der-photovoltaikanlage
[accessed 2 December 2018].

Solarwatt, no year. *Förderung von Stromspeichern.* [Online]
Available at: https://www.solarwatt.de/stromspeicher/foerderung
[accessed 26 November 2018].

sonnen GmbH, 2017. *Die Blockchain ist die nächste Evolutionsstufe der dezentralen Energieversorgung.* [Online]
Available at: https://sonnen.de/wissen/die-blockchain-ist-die-naechste-evolutionsstufe-der-dezentralen-energieversorgung/
[accessed 5 February 2019].

Statista, 2017. *Smart Home.* [Online]
Available at: https://de.statista.com/outlook/279/137/smart-home/deutschland#market-marketDriver
[accessed 21 November 2018].

Straßburg, M., 2018. *Wie wird verkettet? – Wie funktioniert eine Blockchain?.* [Online]
Available at: https://dbcp.online/wie-wird-verkettet-wie-funktioniert-eine-blockchain/
[accessed 16 December 2018].

Stromspiegel, no year. *Die Energiewende - für klimaschonende, zuverlässige und bezahlbare Energie.* [Online]
Available at: https://www.stromspiegel.de/stromkosten/energiewende-strompreise-eeg/
[accessed 10 November 2018].

Swan, M., 2015. *Blockchain: Blueprint for a New Economy.* s.l.:O'REILLY.

SWD, no year. *Digitalisierung der Energiewende.* [Online]
Available at: https://www.swd-ag.de/privatkunden/strom/digitalisierung-der-energiewende/
[accessed 28 December 2018].

Talin, B., 2019. *Blockchain – Möglichkeiten und Anwendungen der Technologie.* [Online]
Available at: https://morethandigital.info/blockchain-moeglichkeiten-und-anwendungen-der-technologie/
[accessed 4 February 2019].

Tatsachen über Deutschland, no year. *Generationenprojekt Energiewende.* [Online]
Available at: https://www.tatsachen-ueber-deutschland.de/de/rubriken/umwelt-klima/generationenprojekt-energiewende
[accessed 7 November 2018].

Thon, S., Utz, M. & Zoerner, T., 2018. *BLOCKCHAIN – EINE TECHNOLOGIE MIT DISRUPTIVEM CHARAKTER.* [Online]
Available at: https://www.vditz.de/fileadmin/media/news/documents/Blockchain_-_Eine_Technologie_mit_disruptivem_Charakter.pdf
[accessed 4 February 2019].

trend:research, 2017. *Eigentümerstruktur: Erneuerbare Energien.* [Online]
Available at: https://www.trendresearch.de/studien/20-01174.pdf?3c1443613a233cd2ad069a2fef34a24d
[accessed 1 December 2018].

Umweltbundesamt, 2013. *Erneuerbare-Energien-Wärmegesetz.* [Online]
Available at: https://www.umweltbundesamt.de/themen/klima-energie/erneuerbare-energien/erneuerbare-energien-waermegesetz
[accessed 8 October 2018].

Umweltbundesamt, 2018a. *Erneuerbare Energien in Zahlen.* [Online]
Available at: https://www.umweltbundesamt.de/themen/klima-energie/erneuerbare-energien/erneuerbare-energien-in-zahlen#statusquo
[accessed 28 November 2018].

Umweltbundesamt, 2018b. *Erneuerbare Energien im Jahr 2017.* [Online]
Available at: https://www.umweltbundesamt.de/themen/erneuerbare-energien-im-jahr-2017
[accessed 30 November 2018].

Umwelt-Energie, 2018. *Digitalisierung: "Die Energiewirtschaft hinkt weit hinterher…".*
[Online]
Available at: https://www.umwelt-energie-report.de/2018/08/digitalisierung-die-energiewirtschaft-hinkt-weit-hinterher.html
[accessed 2 December 2018].

Unit-m, 2018. *Blockchain – Eine Erklärung für Entscheider.* [Online]
Available at: https://www.unit-m.de/blockchain-eine-erklaerung-fuer-entscheider/
[accessed 22 December 2018].

Universität Duisburg-Essen, 2018. *Mittelfristprognose zur deutschlandweiten Stromerzeugung aus EEG geförderten Kraftwerken für die Kalen-derjahre 2019 bis 2023.* [Online]
Available at: https://www.netztransparenz.de/portals/1/Content/EEG-Umlage/EEG-Umlage%202019/20181011_Abschlussbericht%20EWL.pdf
[accessed 17 December 2018].

Unnerstall, T., 2016. *Faktencheck Energierwende. Konzept, Umsetzung, Kosten - Antworten auf die 10 wichtigsten Fragen.* Heidelberg: Springer.

Verband der Elektrotechnik, Elektronik und Informationstechnik, 2007. *VDE-Studie: Dezentrale Energieversorgung 2020.* [Online]
Available at:
https://www.vde.com/resource/blob/792808/db366b86af491989fcd2c6ba6c6f21ad/etg-

studie-dezentrale-energieversorgung2020-komplette-studie-data.pdf
[accessed 27 November 2018].

Voshmgir, S., 2016. *Blockchains, Smart-Contracts und das Dezentrale Web.* [Online]
Available at: https://www.technologiestiftung-berlin.de/fileadmin/daten/media/publikationen/170130_BlockchainStudie.pdf
[accessed 5 January 2019].

Wagenblass, D., 2018. *Photovoltaikanlagen: Was tun nach dem Ende der EEG-Förderung?.* [Online]
Available at: https://partner.mvv.de/blog/photovoltaikanlagen-ende-eeg-foerderung
[accessed 2 March 2019].

Walport, M., 2015. *Distributed Ledger Technology: beyond block chain.* [Online]
Available at: https://assets.publishing.service.gov.uk/government/uploads/system/uploads/attachment_data/file/492972/gs-16-1-distributed-ledger-technology.pdf
[accessed 6 December 2018].

Weniger, J., Bergner, J., Tjaden, T. & Quaschning, V., 2015. *Dezentrale Solarstromspeicher für die Energiewende.* [Online]
Available at: https://speicherinitiative.at/assets/Uploads/06-HTW-Berlin-Solarspeicherstudie.pdf
[accessed 25 November 2018].

Wiegand, J., 2017. *Dezentrale Stromerzeugung als Chance zur Stärkung der Energie-Resilienz.* [Online]
Available at: https://www.econstor.eu/bitstream/10419/162717/1/890878781.pdf
[accessed 22 November 2018].

Wienhold, C., 2016. *Mit Partnern aus Baden-Württemberg die Energiewende 2.0 angehen.* [Online]
Available at: https://www.energie-klimaschutz.de/energiewende-2-0/
[accessed 30 November 2018].

Windmesse, 2018. *Wichtiger Meilenstein für die Digitalisierung der Energiewende: Erstes Zertifikat für Smart-Meter Gateway übergeben.* [Online]
Available at: https://w3.windmesse.de/windenergie/pm/30335-smart-meter-gateway-energiewende-windenergie
[accessed 29 December 2018].

Winkler, R., 2019. *„Blockchain" – Die Lösung für eine dezentrale Energiewende?.* [Online]
Available at: https://blockchain-initiative.de/uncategorized/blockchain-die-loesung-fuer-eine-dezentrale-energiewende/
[accessed 5 January 2019].

Witsch, K., 2018. *Blockchain-Technologie könnte die nächste Energiewende einleiten.*
[Online]
Available at: https://www.handelsblatt.com/unternehmen/energie/strommarkt-blockchain-technologie-koennte-die-naechste-energiewende-einleiten/22837862.html?ticket=ST-720052-li57r5fWfxPnfAvz12aV-ap2
[accessed 4 February 2019].

Wood, G., 2018. *ETHEREUM: A SECURE DECENTRALISED GENERALISED TRANSACTION LEDGER.* [Online]
Available at: https://ethereum.github.io/yellowpaper/paper.pdf
[accessed 24 January 2019].

Yu, A., 2018. *Ethereum Development Tutorial.* [Online]
Available at: https://github.com/ethereum/wiki/wiki/Ethereum-Development-Tutorial
[accessed 24 January 2019].

Zdrallek, M., Uhlig, R., Johae, C. & Harnisch, S., 2016. *Gutachten: Untersuchung des Daten- und Informationsbedarfs der Verteilungs-netzbetreiber zur Wahrnehmung ihres Anteils an der Systemverantwortung.* [Online]
Available at: https://www.bdew.de/media/documents/20161209_Gutachten-Datenbedarf-Verteilnetzbetreiber.pdf
[accessed 2 December 2018].

Zeit, 2018. *Erneuerbare Energie.* [Online]
Available at: https://www.zeit.de/zeit-wissen/2018/04/erneuerbare-energien-stromspeicher-technologie-lithium/seite-3
[accessed 13 November 2018].

Zhaw, no year. *Was ist der Unterschied zwischen Microgrids und Smart Grids?.* [Online]
Available at: https://www.zhaw.ch/de/lsfm/institute-zentren/iunr/oekotechnologien-und-energiesysteme/erneuerbare-energien/microgrids/unterscheidung/
[accessed 27 November 2018].

Zimmermann, H. & Hoppe, J., 2018. *Chancen und Risiken der Blockchain für die Energiewende.* [Online]
Available at: https://germanwatch.org/sites/germanwatch.org/files/publication/21472.pdf
[accessed 28 December 2018].

Zunkunftsenergien, 2019. *„Blockchain" – Die Lösung für eine dezentrale Energiewende?.*
[Online]
Available at: http://www.zukunftsenergien.de/fileadmin/user_upload/zukunftsenergien/Dokumente/PM-19-02-AKZ73.pdf
[accessed 4 February 2019].

Appendix

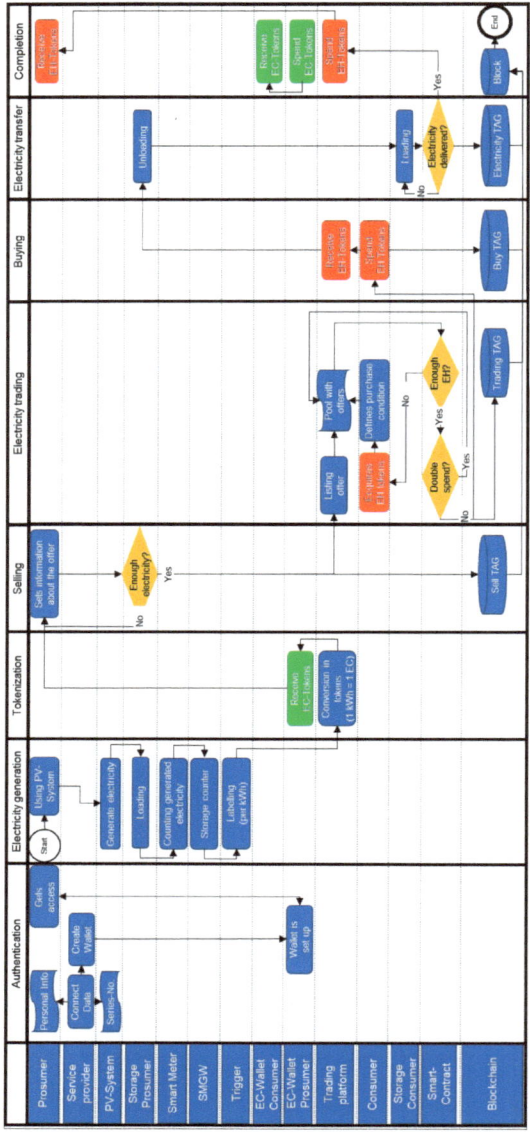

www.ingramcontent.com/pod-product-compliance
Lightning Source LLC
Chambersburg PA
CBHW040315220526
45473CB00009B/2447